Messier
Astrophotography
Reference

Allan Hall

Copyright © 2013 by Allan Hall

10 9 8 7 6 5 4 3 2 1

All rights reserved. No part of this publication may be reproduced, distributed, or transmitted in any form or by any means, including photocopying, recording, or other electronic or mechanical methods, without the prior written permission of the publisher, except in the case of brief quotations embodied in critical reviews and certain other noncommercial uses permitted by copyright law. For permission requests, write to the publisher, addressed "Attention: Permissions Coordinator," at the address below.

Allan Hall
1614 Woodland Lane
Huntsville, TX 77340
www.allans-stuff.com/mar/

Although the author and publisher have made every effort to ensure that the information in this book was correct at press time, the author and publisher do not assume and hereby disclaim any liability to any party for any loss, damage, or disruption caused by errors or omissions, whether such errors or omissions result from negligence, accident, or any other cause.

Any trademarks, service marks, product names or named features are assumed to be the property of their respective owners, and are used only for reference. There is no implied endorsement if we use one of these terms.

All images/graphics/illustrations in this book are copyrighted works by Allan Hall, ALL RIGHTS RESERVED with the exception of the picture of Charles Messier which is in the public domain.

Acknowledgements:

The following persons/companies have graciously agreed to allow reprints of their screens in this publication.

Star charts printed from AstroPlanner V2, used with permission, Paul Rodman, Author

Table of Contents

Getting Started 1

   About the book 1

   Using this book 2

   Charles Messier and his list 7

   The Virgo SuperCluster 9

The Messier Objects 10

   M1:The Crab Nebula 10

   M2 13

   M3 16

   M4 19

   M5 22

   M6:The Butterfly Cluster 25

   M7:The Ptolemy Cluster 28

   M8:The Lagoon Nebula 31

   M9 34

   M10 37

   M11:The Wild Duck Cluster 40

   M12 43

   M13:The Great Globular Cluster In Hercules 46

   M14 49

   M15 52

   M16:The Eagle Nebula 55

   M17:The Omega Nebula 58

   M18 61

M19 64

M20:The Trifid Nebula 67

M21 70

M22:The Sagittarius Cluster 73

M23 76

M24:The Sagittarius Star Cloud 79

M25 82

M26 85

M27:The Dumbbell Nebula 88

M28 91

M29 94

M30 97

M31:The Andromeda Galaxy 100

M32 103

M33:The Triangulum Galaxy 106

M34 109

M35 112

M36 115

M37 118

M38 121

M39 124

M40:The Winnecke 4 Double... 127

M41 130

M42:The Orion Nebula... 133

M43:De Mairan's Nebula 136

M44:The Beehive Cluster 139

M45:The Pleaides 142

M46 145

M47 148

M48 151

M49 154

M50 157

M51:The Whirlpool Galaxy 160

M52 163

M53 166

M54 169

M55 172

M56 175

M57:The Ring Nebula 178

M58 181

M59 184

M60 187

M61 190

M62 193

M63:The Sunflower Galaxy 196

M64:The Black Eye Galaxy 199

M65:The Leo Triplet 202

M66:The Leo Triplet 205

M67 208

M68 211

M69 214

M70 217

M71 220

M72 223

M73 226

M74 229

M75 232

M76:The Little Dumbbell Nebula 235

M77 238

M78 241

M79 244

M80 247

M81:Bode's Galaxy..250

M82:The Cigar Galaxy 253

M83:The Southern Pinwheel Galaxy 256

M84 259

M85 262

M86 265

M87 268

M88 271

M89 274

M90 277

M91 280

M92 283

M93 286

M94 289

M95 292

M96 295

M97:The Owl Nebula 298

M98 301

M99:St. Katherine's Wheel 304

M100 307

M101:The Pinwheel Galaxy 310

M102 313

M103 316

M104:The Sombrero Galaxy 319

M105 322

M106 325

M107 328

M108 331

M109 334

M110 337

Wrapping up **340**

Complete Messier shooting schedule 340

Closing notes 341

About the book

When I first started astrophotography it seemed logical to start shooting some of the brightest and easiest to find targets in the northern hemisphere, the Messier objects. Unfortunately I ran into an interesting problem. I had no idea what I was seeing through the telescope, or on the computer screen. Was that an open cluster or was I not even close to my target?

I went looking for a reference to the Messier objects and I found two different types as related to astrophotography. The first had beautiful images, way beyond anything I would be able to create and so was useless at helping me identify, capture and process my own images. The second was too "snapshot" like and equally useless for me. I needed something midrange and try as I might, I just could not find it.

Several years later I started writing my astrophotography books; *Getting Started: Long Exposure Astrophotography* and *Getting Started: Budget Astrophotography* and it dawned on me that other people may be in the exact same place I was. I can fix that.

The purpose of this book is to be a midrange reference to beginning astrophotographers showing them what the targets look like, where they are, to provide some helpful information on capturing specific targets and what a finished image of the target might look like.

I have also included information on how large the target appears in the field of view and the best time of year to capture it. Lastly, I have attempted to put together a schedule someone can use should they want to capture all 110 Messier objects so that they know what targets are in the sky in what order.

Yes, all of this information is readily available from other sources, but it was valuable for me to have something like this all in one place to use for planning as well as in the field and it is my hope that it will be valuable to you in the same way.

Using this book

For each target you will see a fully processed image of the target followed by a chart running across the page which looks like this:

| Jan | Feb | Mar | Apr | May | Jun | Jul | Aug | Sep | Oct | Nov | Dec |

The area in gray below the months indicates the best time of the year to shoot this particular target from the central US. If you are on the extremes of the east or west coast, you may need to alter your times slightly. I based this off a transit time (transit is the time when it switches from rising to setting) of midnight on the 15th of the month. Earlier months will transit later in the evening, later months transit earlier.

The next thing you will see is the sizing indicators that show approximate sizes of the targets in particular telescopes with a particular camera attached which look like this:

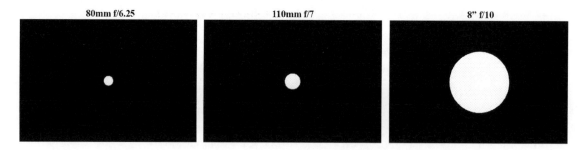

The black rectangles indicate the field of view of an APS-C sized camera sensor (which is roughly the same size as a 25mm eyepiece for reference). I chose the APS-C sensor size as it is the overwhelming choice of beginning astrophotographers.

The round white area is the approximate size the target will appear on your camera sensor, and of course the resultant image without cropping. This is useful both in determining if you are on target correctly and to know what the final image will look like.

I chose the three sizes based off typical astrophotography setups using an 80mm refractor which is extremely common in beginning astrophotography, a 110mm refractor which is common in midrange astrophotography, and the 8" f10 which is a fairly common SCT purchased primarily for visual being re-tasked for astrophotography.

There are targets that fill up or exceed the field of view of the larger telescope, and for this we use the following symbol:

Next will come three sections of text, the first describing the target such as how far away it is, how large it is, how old it is, etc. Second will be the coordinates of the object in right ascension and declination. The third section will contain any notes or tips on how to successfully capture the target.

Lastly, there is a sky chart showing the position of the target in the sky.

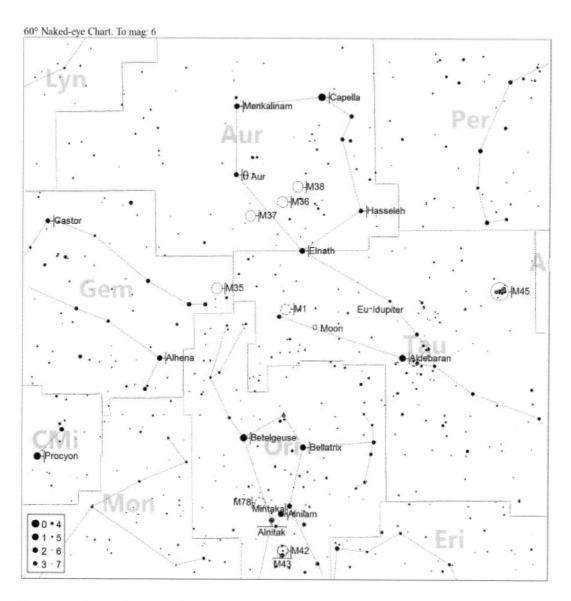

The chart shows the target being discussed directly in the center of the chart, north is up, west is to the right, east to the left, and of course, south is down. Named stars are labeled, and so are other Messier objects in the area. The light gray labels such as Lyn, Aur and Per are constellation abbreviations as shown in the following chart:

And Andromeda	Cru Crux	Oph Ophiuchus
Ant Antlia	Cyg Cygnus	Ori Orion
Aps Apus	Del Delphinus	Pav Pavo
Aqr Aquarius	Dor Dorado	Peg Pegasus
Aql Aquila	Dra Draco	Per Perseus
Ara Ara	Equ Equuleus	Phe Phoenix
Ari Aries	Eri Eridanus	Pic Pictor
Aur Auriga	For Fornax	Psc Pisces
Boo Boötes	Gem Gemini	PsA Piscis Austrinus
Cae Caelum	Gru Grus	Pup Puppis
Cam Camelopardalis	Her Hercules	Pyx Pyxis
Cnc Cancer	Hor Horologium	Ret Reticulum
CVn Canes Venatici	Hya Hydra	Sge Sagitta
CMa Canis Major	Hyi Hydrus	Sgr Sagittarius
CMi Canis Minor	Ind Indus	Sco Scorpius
Cap Capricornus	Lac Lacerta	Scl Sculptor
Car Carina	Leo Leo	Sct Scutum
Cas Cassiopeia	LMi Leo Minor	Ser Serpens
Cen Centaurus	Lep Lepus	Sex Sextans
Cep Cepheus	Lib Libra	Tau Taurus
Cet Cetus	Lup Lupus	Tel Telescopium
Cha Chamaeleon	Lyn Lynx	Tri Triangulum
Cir Circinus	Lyr Lyra	TrA Triangulum Australe
Col Columba	Men Mensa	Tuc Tucana
Com Coma Berenices	Mic Microscopium	UMa Ursa Major
CrA Corona Austrina	Mon Monoceros	UMi Ursa Minor
CrB Corona Borealis	Mus Musca	Vel Vela
Crv Corvus	Nor Norma	Vir Virgo
Crt Crater	Oct Octans	Vol Volans
		Vul Vulpecula

The boxes around the constellations are the constellation boundaries.

Down in the bottom left is a key that shows how a star will be displayed depending on its magnitude. The larger the star is drawn, the brighter it is, and the lower the magnitude number.

These charts are printed from AstroPlanner by iLanga Inc which is available from www.astroplanner.net. If you are at all interested in software planning, observation logging and charting software, you owe it to yourself to take a look at this software. To me it seems to be far and away the best software for the job. When you do visit his site, be sure to thank the author, Paul Rodman, for being kind enough to allow me to reprint charts made with his fine software.

Objects do not have a "correct" exposure as the exposure needed will vary with the amount of light pollution in the area, sensitivity of your equipment (CCDs are typically more sensitive than a DSLR for example, monochrome is generally more sensitive than color, etc), focal ratio of your telescope (an f5 scope lets in more light than an f8 scope), and more. The exposure information given in this book is a reference or starting point only. If you find that on one target my exposures are far too short, you may need to increase all the exposure numbers in the book to help match your equipment and skies. When I give example exposures, they are for my 110mm f7 APO refractor with my Nikon D7000 camera; modify them accordingly.

Throughout the book you will hear me talk about different methods of processing and specific terms such as levels, curves, white points, black points, layers and more. Unfortunately it is beyond the scope of this book to include detailed processing examples and a huge glossary. For that information you may want to acquire one of my other books such as *Getting Started: Long Exposure Astrophotography* or *Getting Started: Budget Astrophotography*.

There are a few terms you need to understand before getting into the objects:

Core – The central section of an object where it is densest.

Deep – Generally referring to longer exposures as these will capture objects "deeper" in space.

Field – Usually referring to the background but means the entire area of the image.

Spread – The area around the core of a globular cluster, usually much fainter than the core and much less dense.

Charles Messier and his list

Charles Messier was born in Badonviller France just north of Lac de Pierre-Percée and about 80km west of Strasbourg France on the 26th of June, 1730. Despite a difficult childhood where six of his eleven siblings and his father all passed away before he was twelve, Charles became fascinated by astronomy. Much of this interest could be attributed to two celestial events in his youth: the appearance of the "Great Comet of 1744" (C/1743 X1) and the annular solar eclipse of the 25th of June 1748. Unfortunately when Charles' father passed away, when Charles was eleven, he was forced to leave school and continue his education at home with his brother as instructor.

When he was twenty one years old, in 1751, Charles began working for French Naval Astronomer, Joseph Nicholas Delisle, who was originally instructed by Jacques Cassini and had overseen the building of a new observatory in the palace of Cluny in 1747. This sped his advancement in astronomy such that in eight short years he became chief astronomer of the Marine Observatory and in thirteen years he was made a fellow of the Royal Society.

Charles' primary focus was on comets, of which he discovered thirteen. He is most famous however for his list of bright objects in the skies of the northern hemisphere. This came about because on the 28th of August 1758 while he was searching the skies for comet Halley he noticed a stationary (as compared to the stars) fuzzy object in the Taurus constellation. Charles noted this object so that it would not be confused with a possible comet sighting as he worked. This object went on to become known as Messier 1, M1, or the Crab Nebula. Over the years he cataloged over one hundred objects in his list, and later on researchers added a few that he had noted but not put on the list to bring the total to one hundred and ten.

Both M1 and M2 were objects that had been previously discovered, by English astronomer John Bevis in 1731 and Italian astronomer Giovanni Domenico Maraldi in 1746 respectively. Charles' first true discovery was the globular cluster M3 in the Canes Venatici constellation on the 3rd of May 1764.

Charles passed away on the 11th of April 1817 at the age of 87 and was reported to have participated in astronomy virtually up to the end of his life.

The list that bears his name continues today as the single most recognizable list in astronomy, probably due to the fact that it is relatively easy under dark skies to observe all of the Messier objects with nothing more than a reasonable set of binoculars. Many of the targets are quite visible with the naked eye such as M31, M42 and M45.

The Virgo "SuperCluster"

While this really is not directly related to the Messier list, it is important to understand because it is mentioned several times throughout the book. In the constellation of Virgo there is a high concentration of galaxies that no matter how hard I try to describe, you just have to see to grasp:

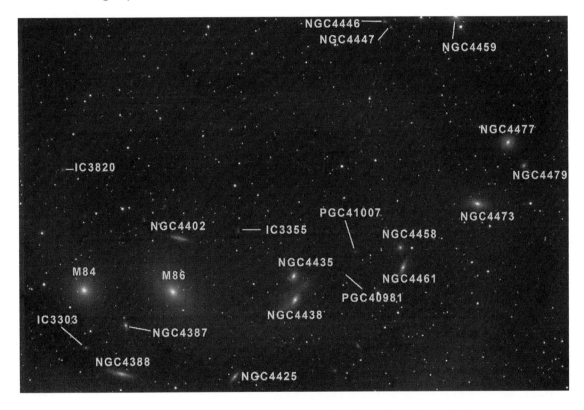

This is the area around M86, and this is one single shot through a 110mm refractor telescope, not a mosaic or large part of the sky. Every M#, NGC#, IC# and PGC# you see is a galaxy. This particular area is called "the chain".

Some catalogs call this the Virgo Cluster, others call it the supercluster, and hopefully you see why I choose the latter.

Messier 1: The Crab Nebula

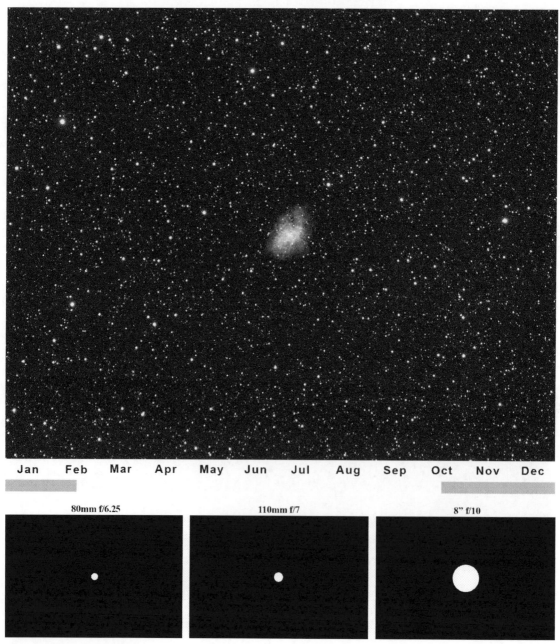

About the target: This supernova remnant in Taurus spans 11 light years across and sits some 6,500 light years away. At the center is a pulsar which is the remnant of the star that went supernova in 1054 as reported by Chinese astronomers. Although this is the first object in Charles Messier's list, it was actually discovered in 1731 by English astronomer John Bevis.

Where it is:
Right ascension: 05h 34m 31.94s
Declination: +22° 00′ 52.2″

Imaging the target: The Crab Nebula is a pretty small target as far as nebulas go, in addition to that it is right in some pretty heavy light pollution from my dark site, making it pretty difficult to grab good images. There is a surprising amount of detail in this little nebula, especially around the edges. I found that this target really benefits from extremely good polar alignment and guiding. The images I did capture at ISO 800 for 240 seconds each (any longer and the sky fog became horrible, even with a light pollution filter) appeared pretty good.

Since this is such a detailed small object it is more important than normal to shoot it in cold weather, or with a camera that can be cooled quite a bit. Thermal noise can strip out the details from the edge and center of the nebula very easily. This becomes a little less critical if you are shooting with a longer focal length scope such as an 8" SCT but that presents its own issues with more difficult guiding.

The good news is that the nebula does not have that much brightness differential between the core and the outer edges so a single set of exposures should be able to get maximum detail.

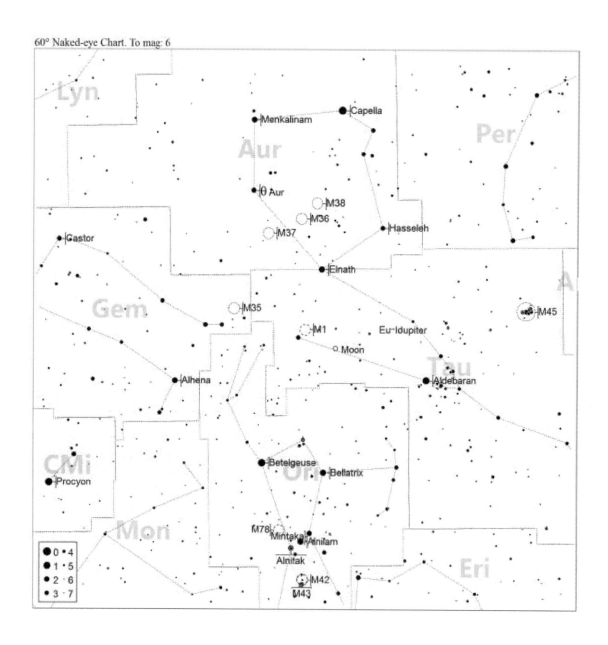

Messier 2:

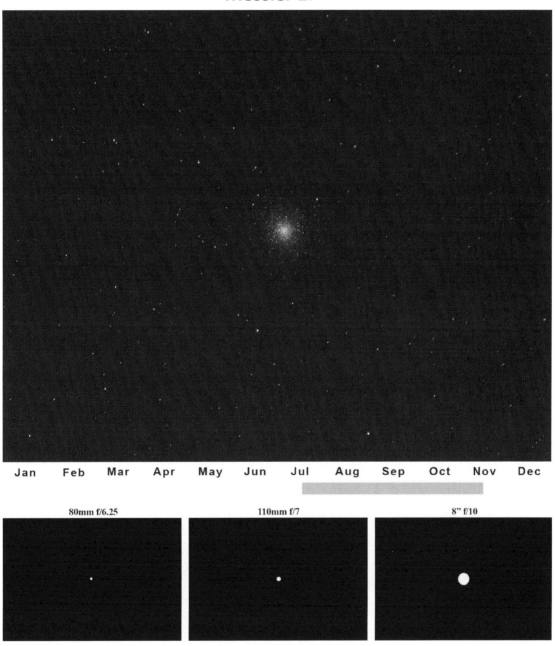

About the target: Discovered around 1746 in the constellation Aquarius by Italian/French astronomer Jean-Dominique Maraldi who originally thought it to be a nebula. It was eventually verified to be a cluster by William Herschel in 1783. This is one of the largest globular clusters in the sky at about 175 light years across and a little over 37,000 light years away. The cluster contains roughly 150,000 stars.

Where it is:
Right ascension: 21h 33m 27.02s
Declination: −00° 49′ 23.7″

Imaging the target: This globular cluster has a reasonably tight core and also a decent spread so it will be difficult to capture and stretch with a single set of exposures. The background of stars certainly can use a long exposure to capture although I probably would not go beyond 400 or so seconds with my f7 refractor at 800 ISO due to a few fairly bright stars in the field that can easily be blown out.

I would probably recommend three sets of images, each spaced a full stop apart. What this means is that if I shot 20 images at ISO 800 and 200 seconds, I would then shoot a second set of 20 images at 400 seconds and a third at 100 seconds, all at the same ISO 800. These could be processed individually and then stacked in your choice of HDR software or using layers in your image editing package.

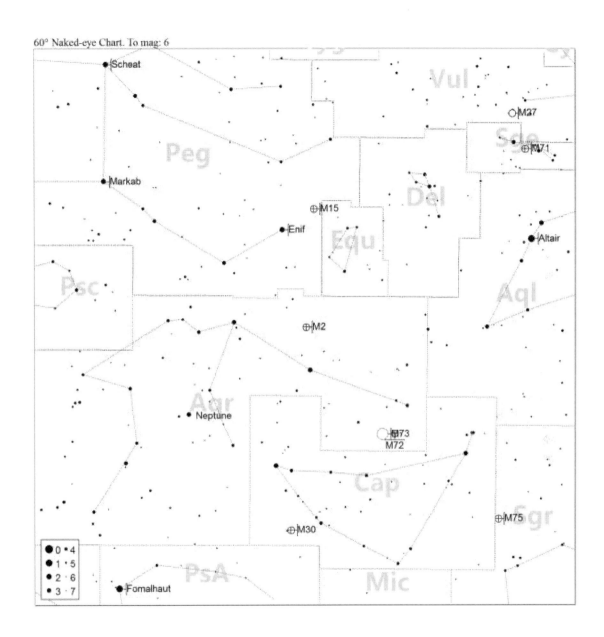

Messier 3:

About the target: Discovered on May 3rd, 1764 by Charles Messier, this globular cluster sits in the constellation Canes Venatici, is some 33,000 light years away and spans some 180 light years across. There are 274 identified stars in the cluster and is easily viewable in small telescopes. This is considered one of the finest globular clusters to view in the northern hemisphere.

Where it is:
Right ascension: 13h 42m 11.62s
Declination: +28° 22' 38.2"

Imaging the target: This is a prime example of a globular cluster that really requires multiple sets of exposures to capture well. The image you see for this target is a combination of 10 images at 30 seconds, 10 images at 100 seconds, 10 images at 300 seconds and 10 images at 600 seconds all shot at ISO 800. In retrospect I should have stretched the 600 second images more to reveal more background stars and masked out the central region so as to not blow it out.

Still, this shows that the central core can be preserved fairly well while still getting some of the outlying area. The reduced stretching I gave this might have omitted some of the background we would like to have but it also made sure there was little to no noise as the black of space appears perfectly smooth and noiseless.

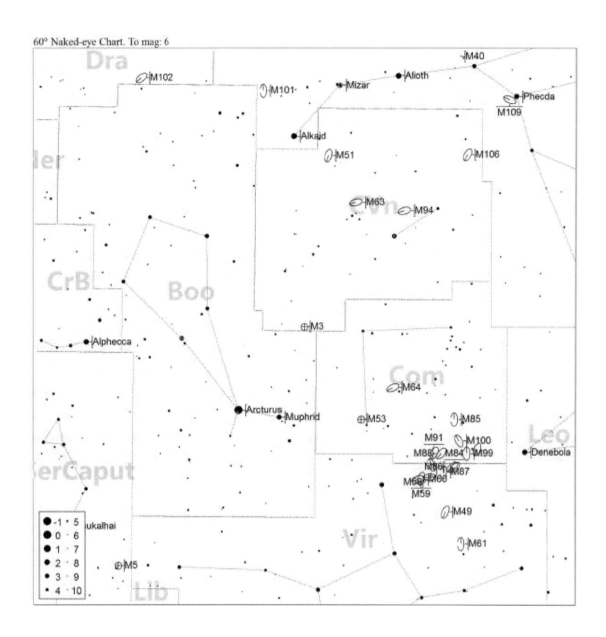

Messier 4:

About the target: Discovered in 1746 by Swiss astronomer Jean-Philippe de Chéseaux this globular cluster is approximately 7,200 light years away in the constellation of Scorpius. Spanning some 70 light years it contains some of the oldest known stars in our galaxy at 13 billion years old. According to spectroscopic observations this cluster appears to be two distinct clusters, or one cluster that has gone through two cycles of star formation.

Where it is:
Right ascension: 16h 23m 35.22s
Declination: −26° 31′ 32.7″

Imaging the target: This globular is one of the few that can be captured pretty well with one set of exposures if you watch your stretching due to the fact that its core is not as dense as many others. The problem I have with this image is that the stretching combined with the warming temperatures when I shot it resulted in some severe noise. To combat this I plan on shooting this again and shooting at least three times the amount of frames I shot last time (10 240 second images). Noise reduction software is not a good option with this because of all the faint stars in the background which could easily and undesirably be removed. More time on this target would also help with star coloring if you are shooting color images.

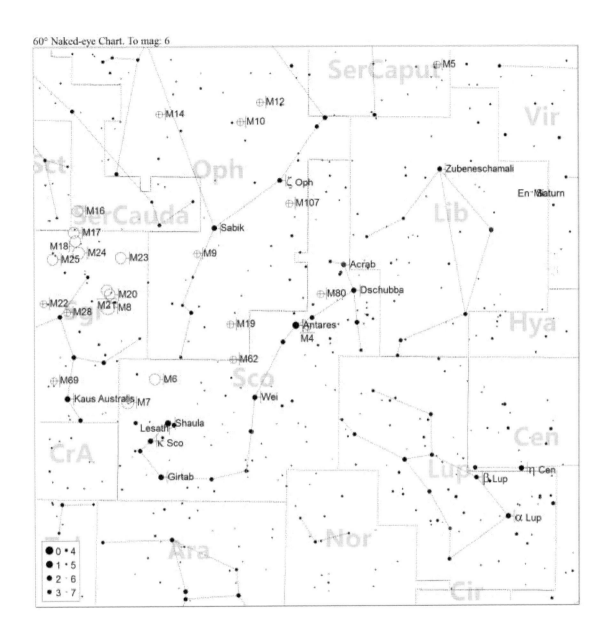

Messier 5:

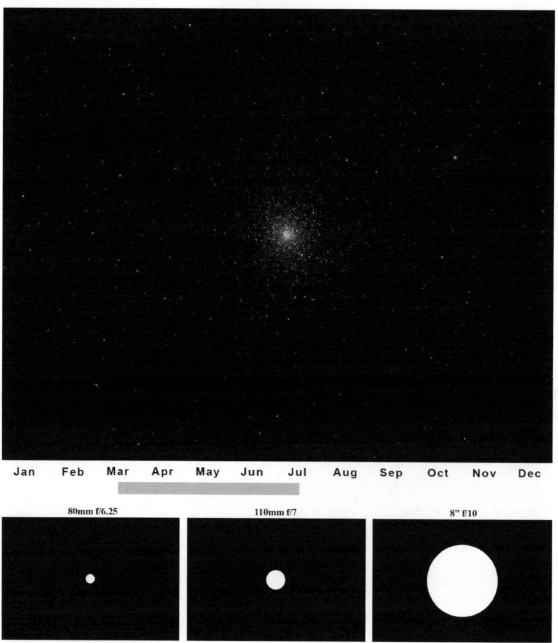

About the target: This globular cluster in Serpens was discovered by German astronomer Gottfried Kirch in 1702 and spans some 160 light years across at a distance of almost 25,000 light years. Along with M4, this is one of the oldest known globular clusters at 13 billion years old. In 1791 William Herschel counted nearly 200 individual stars in this massive cluster. An interesting note is that there are over 100 known variable stars in this cluster.

Where it is:
Right ascension: 15h 18m 33.22s
Declination: +02° 04' 51.7"

Imaging the target: Here is another example of a globular that might benefit from having multiple sets of exposures combined using HDR or masking. The first problem is the dense and bright central core combined with that bright star in the upper right. This does not work well with the faint stars on the outer edge of the cluster. In the image for this target I attempted to use shorter exposures (150 seconds at ISO 800) and moderate gray point stretching which almost worked. The result was that the core is less defined than I would like but not completely blown out, and the star in the upper right is moderately bloated. Unfortunately there is also no real star color in the color version either.

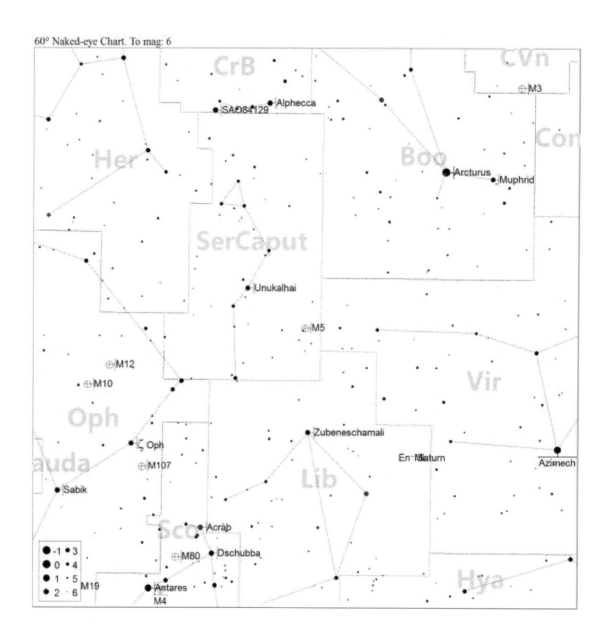

Messier 6: The Butterfly Cluster

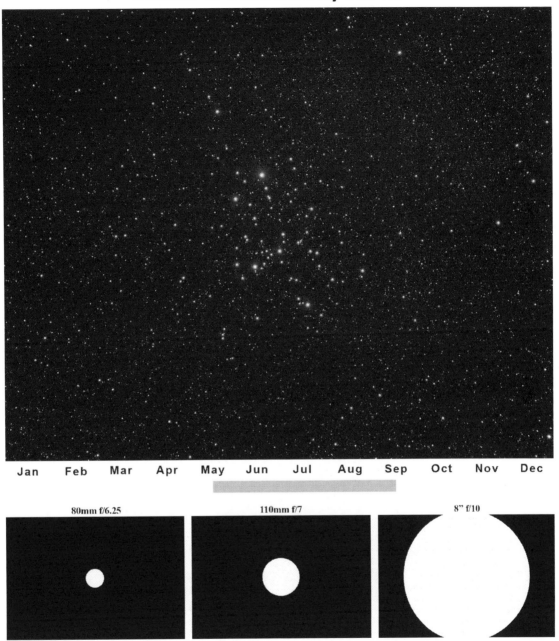

About the target: The butterfly cluster. Robert Burnham JR of Burnham's Celestial Handbook suggests that this cluster may have been discovered in the 1st century by Ptolemy due to its close proximity to the Ptolemy Cluster. The first recorded observation of this cluster came in 1654 by Italian astronomer Giovanni Battista Hodierna. The cluster has a diameter of approximately 12 light years and is about 1,600 light years away in the constellation of Scorpius. Images will show a preponderance of blue stars with one outstanding orange giant dominating the rest.

Where it is:
Right ascension: 17h 40.1m
Declination: −32° 13′

Imaging the target: The butterfly cluster is a nice open cluster that is fairly easy to capture. The only real issue here is that you can blow out some of the brighter stars trying to get a background full of fainter stars. If you can resist that urge you will be rewarded with a beautiful set of blue stars with one very bright yellow star standing out on one side of the cluster. Since this cluster is up in the warmer months you will doubly benefit from shorter exposures as the sensor will not heat up as much. I would however recommend at least 20-25 exposures and plenty of darks to keep the thermal noise in check and provide for a smooth black background. For this target I shot 240 seconds at ISO 800.

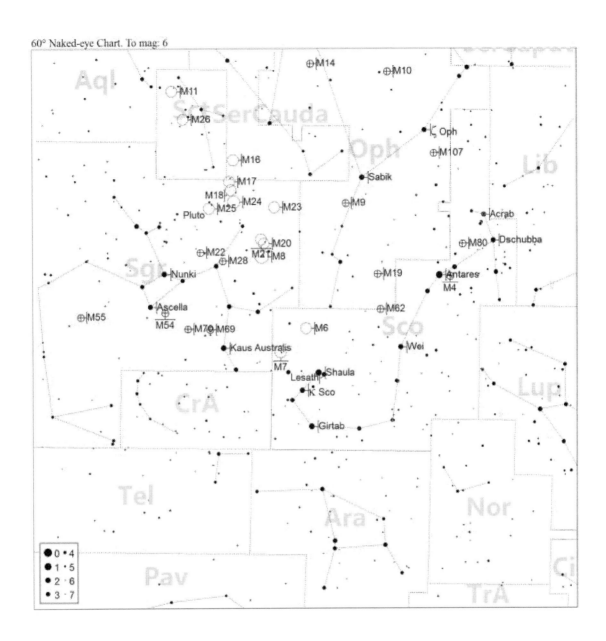

Messier 7: Ptolemy Cluster

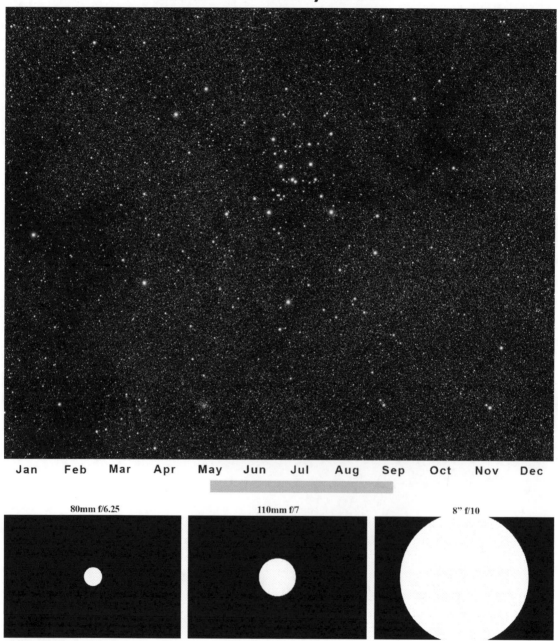

About the target: At just under 1,000 light years away and spanning some 50 light years across, this 200 million year old cluster in Scorpius is easily viewable with no optical aids under moderately dark skies. First recorded by Ptolemy in 130 AD he called it a nebula. It was not known as a cluster until Italian astronomer Giovanni Batista Hodierna in 1645 counted approximately 30 individual stars grouped together. Today it is known as the Ptolemy Cluster and we know it contains some 80 stars.

Where it is:
Right ascension: 17h 53m 51.2s
Declination: −34° 47′ 34″

Imaging the target: This is almost the perfect open cluster to shoot as it provides an excellent field of beautiful blue stars, one standout white/yellow, and a wonderful rich background of stars even at shorter exposures. This is another target where I settled on 240 second exposures at ISO 800, and another target in the warmer months so be sure to take 20-25 exposures and 25 or so darks to keep the thermal noise in check. The thermal noise is particularly important here with the incredible field of background stars that can really turn to sand if you take too few exposures.

Another interesting thing that could potentially be pulled out of this target are the dark lanes in the upper right, through the center area and then again on the lower left. This is an excellent target to really make those dark lanes pop out but you will need a lot of exposures to do that.

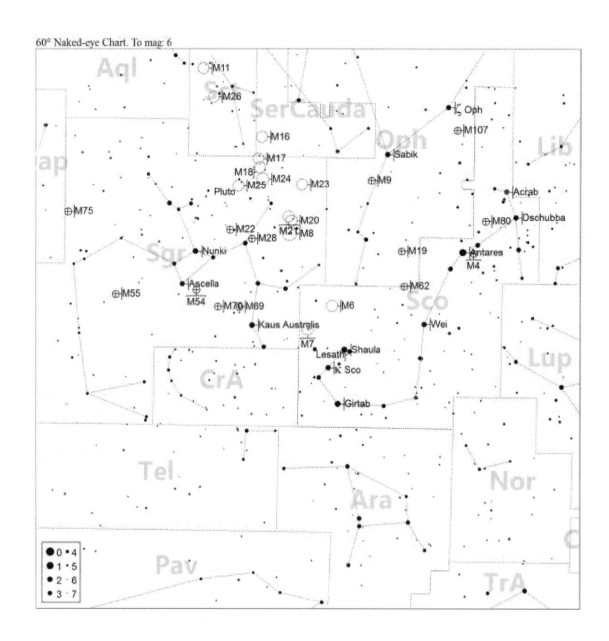

Messier 8: The Lagoon Nebula

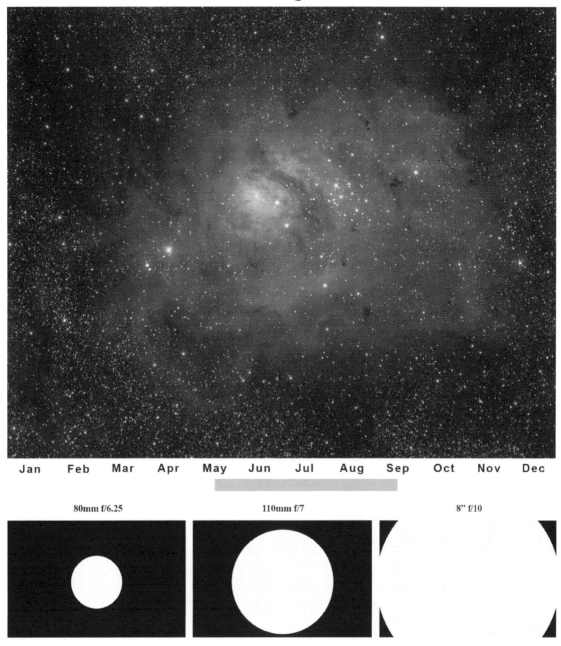

Jan Feb Mar Apr May Jun Jul Aug Sep Oct Nov Dec

80mm f/6.25   110mm f/7   8" f/10

About the target: Discovered by Guillaume Le Gentil in 1747 this star forming emission nebula is one of only two which can be seen with the naked eye from a northern latitude. At over 100 light years in diameter and being around 5,000 light years distant it is a fairly large object in the sky and sits in the constellation Sagittarius. With a reasonable pair of binoculars or small telescope, nebulosity is readily apparent.

Where it is:
Right ascension: 18h 03m 37s
Declination: −24° 23′ 12″

Imaging the target: This is by far one of my favorite objects to image. It is also one of the first nebulas I ever shot. You can do a very good job with this target with only one set of exposures but as with most nebulae in the Messier list, you can get a lot more detail with more shots. That amount of detail can begin to get really impressive. My image shown of this target is 36 images at 300 seconds each at ISO 800, and it could use more. The central region is really popping with detail if you look close.

Again this is a warmer month target so make sure you take lots of darks to help counter that thermal noise.

If you are shooting color here, this nebula is much like M16 and M17 in that it is not bright red as many people seem to think. In regards to M17 for example, according to the *Students for the Exploration and Development of Space*, "The color of the Omega Nebula is reddish, with some graduation to pink. This color comes from the hot hydrogen gas which is excited to shine by the hottest stars which have just formed within the nebula. However, the brightest region is actually of white color, not overexposed as one might think." So don't spend a lot of time trying to make it redder than it appears, unless of course you just want to.

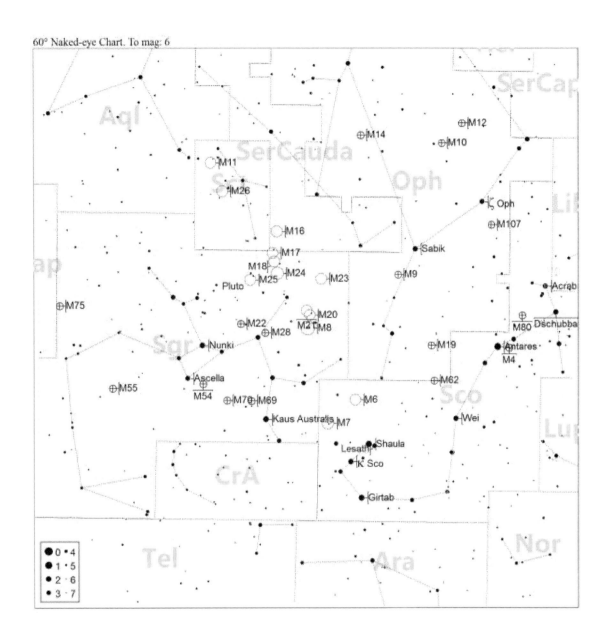

Messier 9:

About the target: In the constellation of Ophiuchus some 5,500 light years away sits this globular cluster. It spans some 90 light years across and was one of many objects discovered by Charles Messier who cataloged it on the 28th of May, 1764. This cluster is dimmer than one might expect as it is sitting right on the edge of Barnard 64, a large dark nebula, and therefore is partially obscured by the dust. An interesting note is that M9 is moving away from us at an extremely high velocity, approximately 224 km/sec.

Where it is:
Right ascension: 17h 19m 11.78s
Declination: −18° 30′ 58.5″

Imaging the target: This is a very interesting globular cluster for me as it has a nice tight core but that core is well defined, it also has a nice little spread coming out a ways, it is small but still large enough to capture well, and lastly, there is that whole area below and to the right in the image that appears to be obscured by a large cloud of dust. Even at a single set of exposures this target comes out relatively well, keeping the number of images high and the exposure time low (25 or so images of ISO 800 240 seconds on my setup would do nicely) to keep the thermal noise to a minimum since this is a warm weather target.

At some point in time I would really like to shoot a large set (50 images or so) at very long exposure times (480 sec or so) just to see what I can drag out of that cloud of dust. I could then use HDR techniques or layering to merge the two sets together and possibly get something really interesting.

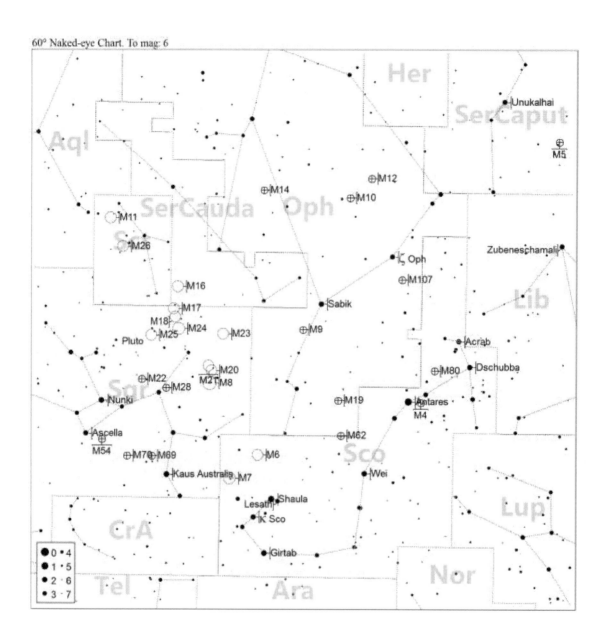

Messier 10:

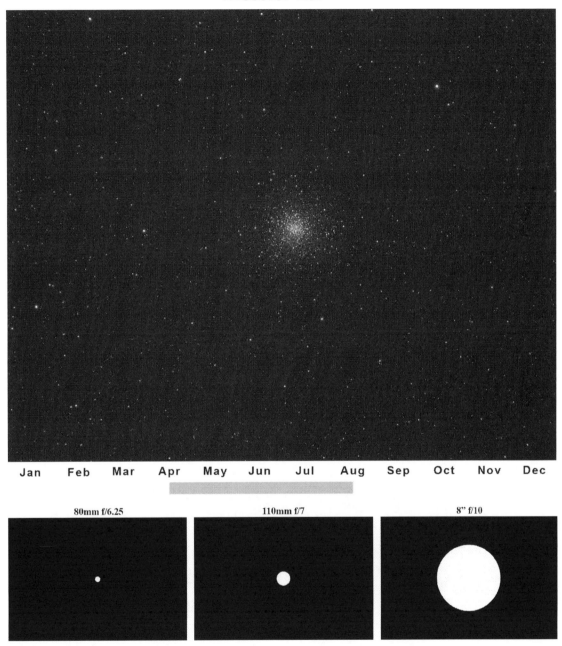

About the target: Discovered by Charles Messier on the 29th of May, 1764 in the constellation of Ophiuchus this globular cluster is some 14,000 light years away and spans approximately 83 light years in diameter. It is one of the brightest globular clusters in this portion of the sky and with larger telescopes can be seen to be as large as 67% the size of a full moon. M10 is known for having an unusually low number of variable stars, reported by some reference sources as only four.

Where it is:
Right ascension: 16h 57m 8.92s
Declination: −04° 05′ 58.07″

Imaging the target: This globular really wanted to blow out the central core on me so I was forced to use very short exposures (150 seconds at ISO 800) to keep that from happening. Unfortunately that had the effect of leaving me with a fairly bland background which could have been much deeper, and a spread that also was very restricted. This target just screams to be redone with two or more sets of exposures and merged together.

Again this is a warm weather target and so you should take many shots and many darks (20-25 of each) to help keep the thermal noise down. I used 20 lights and 24 darks with minimal stretching and it seems to have done a reasonable job.

60° Naked-eye Chart. To mag: 6

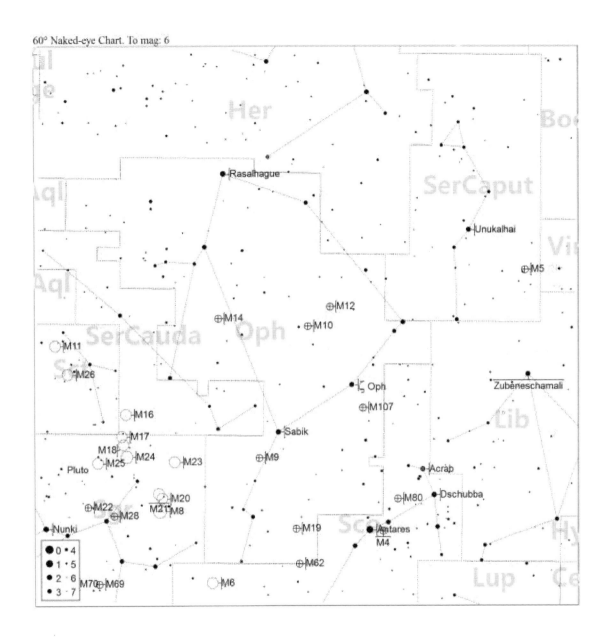

Messier 11: The Wild Duck Cluster

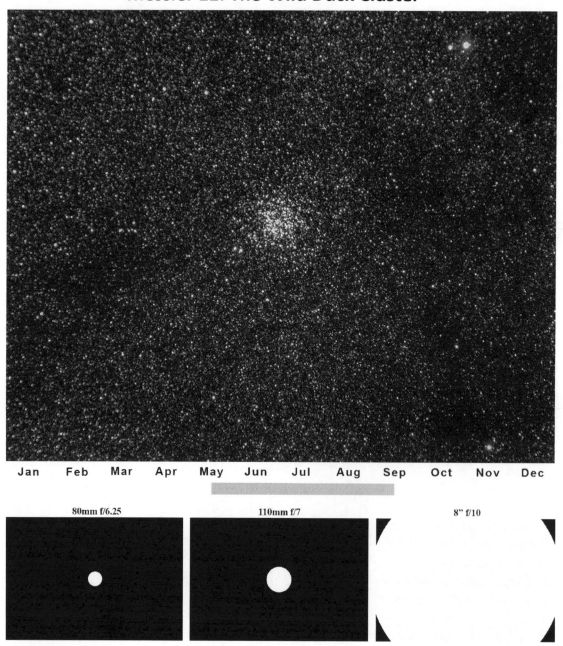

About the target: Sitting in the constellation Scutum is this colorful open cluster discovered in 1681 by the German astronomer Gottfried Kirch. This is one of the most spectacular open clusters in the sky sporting almost 3,000 stars. An open cluster is a group of stars that were formed at roughly the same time in the same molecular cloud. This one is estimated to be over 200 million years old. Messier described the cluster in 1764 as "Cluster of a great number of small stars".

Where it is:
Right ascension: 18h 51.1m
Declination: −06° 16′

Imaging the target: The Wild Duck Cluster is a fantastic target to play with. You absolutely can do some excellent work with only one set of exposures, or you can really mix it up, do multiple sets and have something really special. My shot is a single set of images where I shot 30 100 second images at ISO 800, very short for me, and then stacked them and did a minimum of stretching. The colors really jumped out with this approach however I did bloat the central stars more than I would have liked. The bloating on this image is also partially due to the light pollution filter I was using at the time and has since been replaced so I would be interested in seeing what my results would be like now.

Depending on how you frame the target be careful stretching it. There are a few fairly bright stars shown in my image in the upper and lower right side that can really get away from you when stretching.

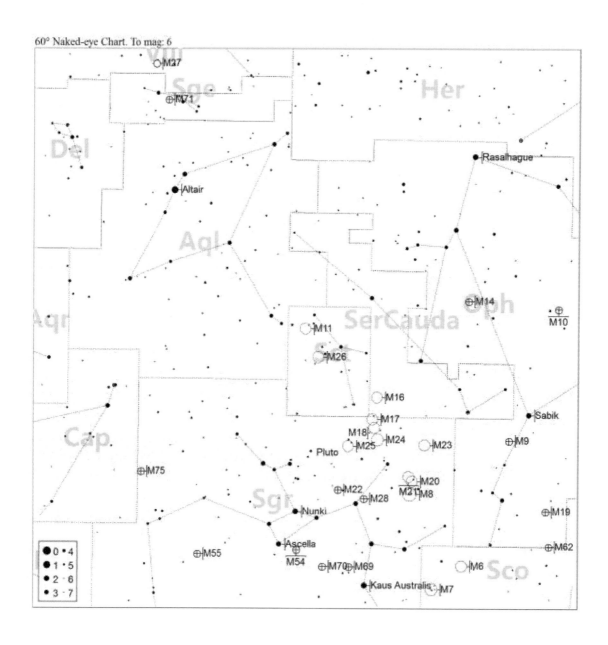

Messier 12:

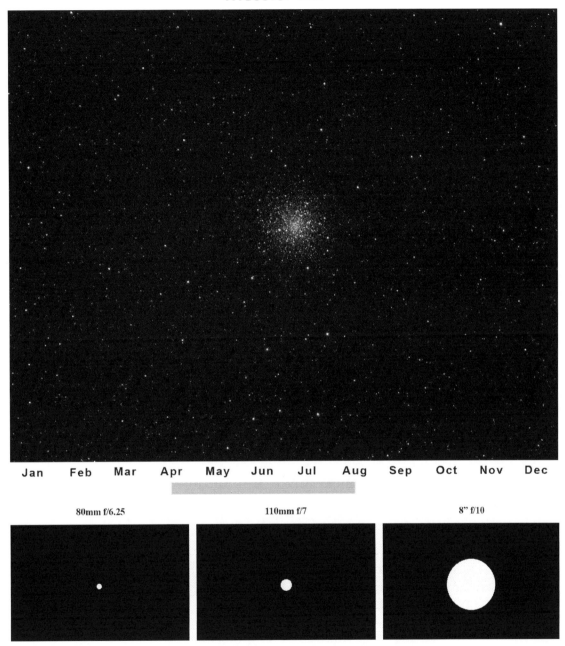

Jan Feb Mar Apr May Jun Jul Aug Sep Oct Nov Dec

80mm f/6.25   110mm f/7   8" f/10

About the target: Discovered by Charles Messier on the 30th of May, 1764, the day following the discovery of M10, most likely because they are only separated by 3 degrees in the sky. Located in the constellation of Ophiuchus just under 16,000 light years away and having a diameter of approximately 74 light years, this cluster can be seen with only a small telescope or pair of binoculars.

Where it is:
Right ascension: 16h 47m 14.18s
Declination: −01° 56′ 54.7″

Imaging the target: This loosely packed globular cluster could use a little more exposure than what my image shows (150 sec at ISO 1600). I was trying to keep the exposure time down since this was June in Texas and that means it was pretty dang hot so I wanted to minimize thermal noise. That did work, but the loss of dynamic range and increase in sensor noise pretty much negated everything I gained. If I were to reshoot this I would try about 300 seconds at ISO 800, a fairly deep exposure, so I could pull out the spread a little more and hopefully a bit more of the background stars. Even the central core in this shot could be brighter without hurting the overall image. Looking close at the image I can see it has a pretty impressive spread and I would really like that to pop out more in the image.

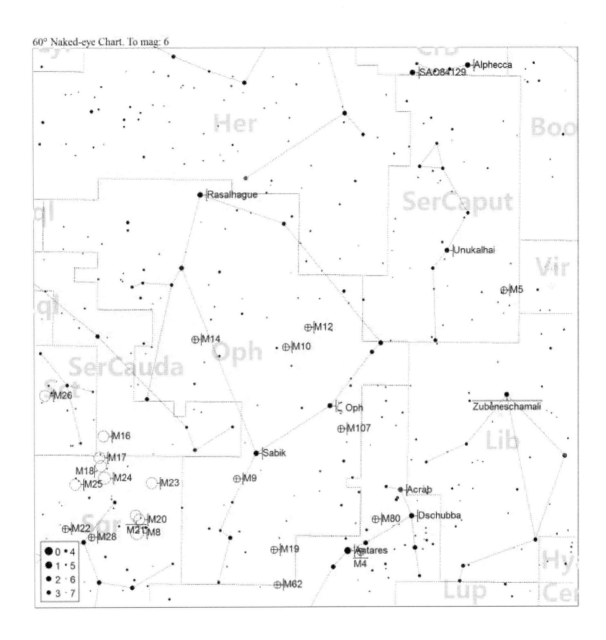

Messier 13: The Great Globular Cluster In Hercules

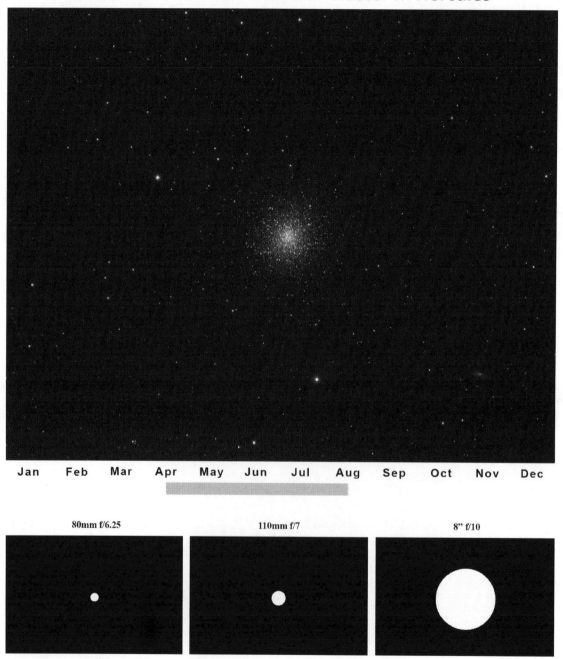

About the target: If this looks like a big glob in the sky there is a good reason for this, it is a globular cluster of some 300,000 stars in the constellation of Hercules. Discovered in 1714 by English astronomer Edmond Halley (who also computed the orbit of Halley's comet) it is over 25,000 light years from Earth and is about 145 light years in diameter. A globular cluster is a group of stars which orbits a galactic core and is bound by gravity, giving it the roughly spherical shape.

Where it is:
Right ascension: 16h 41m 41.24s
Declination: +36° 27' 35.5"

Imaging the target: This is one of the most impressive globular clusters in the sky, if not the most impressive. It has an easily defined core and a wonderful spread, much nicer than my image shows. It is also one of the most imaged targets in the sky, so I put most of my time on other, less imaged targets.

Now that is not to say you cannot do amazing things with this one. Even with a single set of exposures the core and spread can both be excellently represented. I shot a medium depth exposure of 240 seconds at ISO 800 for this image. I did shoot a lot of darks, flats and bias files (25 of each) to make sure I got rid of all the thermal noise and other defects I could since this was shot in July in Texas, so I was sweating at midnight it was so hot.

The background for M13 is not as impressive as I would have liked, so you might consider taking some longer exposures to layer together with the normal images to get a more impressive field of stars in the background. If you happen to have a long focal length scope such as an 8" SCT this target can be very interesting as it takes up most of your field of view. Using a scope like an 8" SCT however will almost require you to use multiple sets of exposures to define both the core and spread.

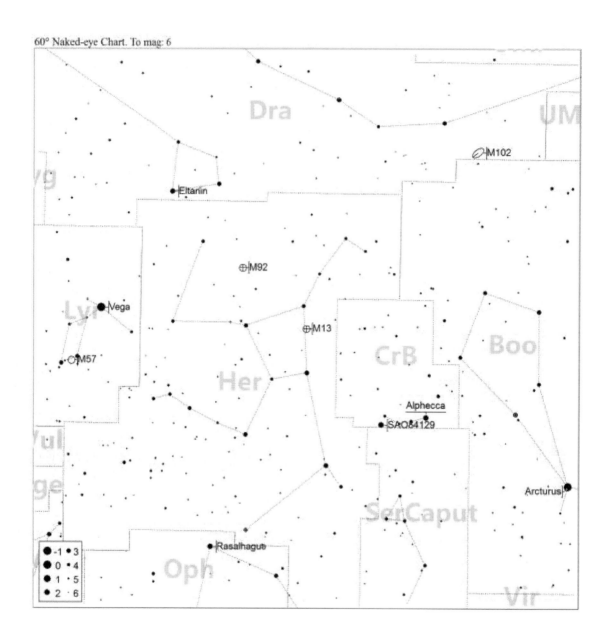

Messier 14:

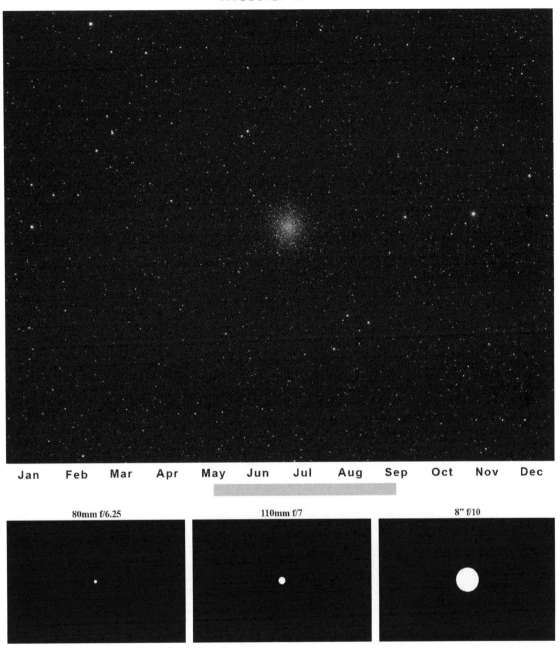

About the target: Discovered by Charles Messier on the 1st of June, 1762 this globular cluster in Ophiuchus is said to contain several hundreds of thousands of stars. Sitting some 30,000 light years away and spanning almost 100 light years, it is a massive globular even though it appears smaller and dimmer than its neighbors M10 and M12. In 1938 photographic evidence was obtained of a nova in the cluster, one of only two known novae in globular clusters.

Where it is:
Right ascension: 17h 37m 36.15s
Declination: −03° 14′ 45.3″

Imaging the target: Though not as impressive as something like M13, M14 is a lot easier to shoot. With one set of exposures you can achieve a reasonable background of stars, and a reasonably well defined core. The only issue you will have with this approach is the spread which is pretty dim as shown in the image.

Another issue is that there are some pretty bright stars on the right center, left top, and left center which can be fairly easy to blow out if you try too much exposure and/or stretching. Fortunately these are pretty far out from the target so a little cropping might serve you well. Unless I had a real reason to push hard on this image I would probably just use one set of images and leave it as is, just like the picture shown in this book for this target. If it were a more impressive target like M13 I might get a little more excited.

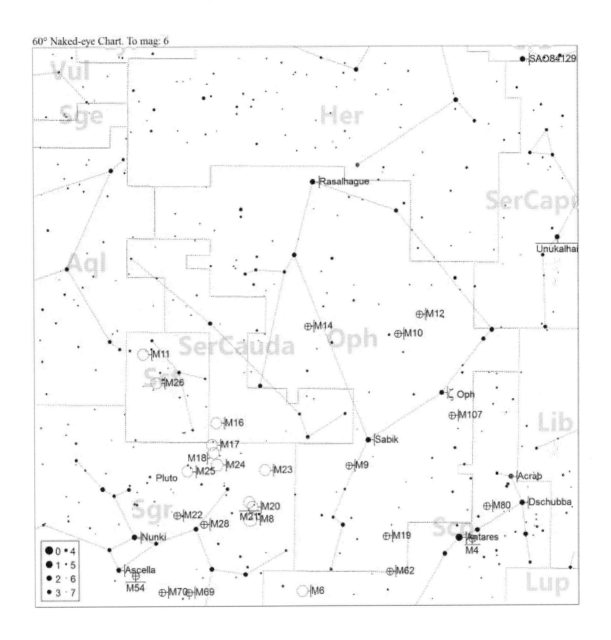

Messier 15:

About the target: M15 was discovered in 1746 by the Italian born French astronomer Jean-Dominique Maraldi. It sits over 30,000 light-years from Earth and is approximately 175 light-years in diameter. This is one of the most densely packed clusters in our galaxy, and is also one of the oldest known at approximately 12 billion years old. It is suspected that this cluster houses over 100,000 stars.

Where it is:
Right ascension: 21h 29m 58.33s
Declination: +12° 10′ 01.2″

Imaging the target: M15 is a pretty large cluster, and fairly bright, so short exposures work pretty well. What you see here is only about twelve 100 second exposures at ISO 800 and the results are not terrible. I certainly could have used more frames and could have used a second set of longer exposures to pull in more stars for the background. With that said, the core is not without structure and the spread is captured fairly well so overall I am pleased with this quick run.

In processing I never touched the white point, preferring instead to stretch only the midpoint. Once the initial stretch was done, I pulled the black down a little to give it more of the inky black I prefer. This of course did darken a few of the really dim stars in the field but I felt this was a worthwhile trade.

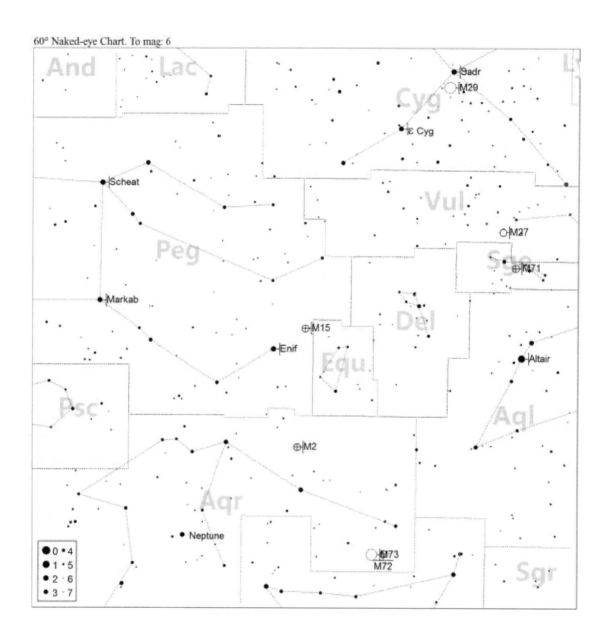

Messier 16: The Eagle Nebula

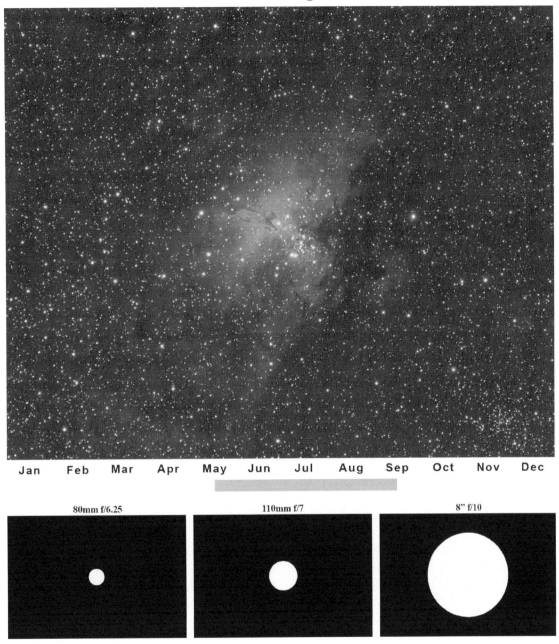

About the target: The Eagle Nebula is an emission nebula in the Constellation Serpens discovered by Swiss astronomer Jean-Philippe de Chéseaux in 1745 and contains an open cluster of stars and the famous "Pillars of Creation" which is the region in the center (to the back of the Eagle's head as shown in the previous image) of star forming gas and dust. The nebula is about 6,500 light-years away.

Where it is:
Right ascension: 18h 18m 48s
Declination: −13° 49′

Imaging the target: This is just an awesome target, but to pull off a really great image you need to make sure your polar alignment, focus and exposures are all dead on. The reason is that most people love a few specific things about this target: 1) The colors in the stars in the field: from bright blue to white and all the way to a wonderful red/orange, the colors are stunning. 2) The amount of nebula: there is more than my image shows but with anything, it is a balancing act and I feel my image shows an "appropriate amount" without getting stretched too hard. 3) The detail in the "pillars of creation" portion of the image which has been made famous by the Hubble Telescope images of this target.

For this target you will want to shoot deep. I shot 480 seconds at ISO 800, and you need 16 or more lights to start with. Since it is just starting to get rather warm when this target is up, thermal noise is pretty easy to deal with. When stretching, you really want to leave your white point alone and work with curves so you can get the best stretch you can without destroying either your star colors or noise levels.

If you are shooting color here, this nebula is much like M8 and M17 in that it is not bright red as many people seem to think. In regards to M17 for example, according to the *Students for the Exploration and Development of Space*, "The color of the Omega Nebula is reddish, with some graduation to pink. This color comes from the hot hydrogen gas which is excited to shine by the hottest stars which have just formed within the nebula. However, the brightest region is actually of white color, not overexposed as one might think." So don't spend a lot of time trying to make it redder than it appears, unless of course you just want to.

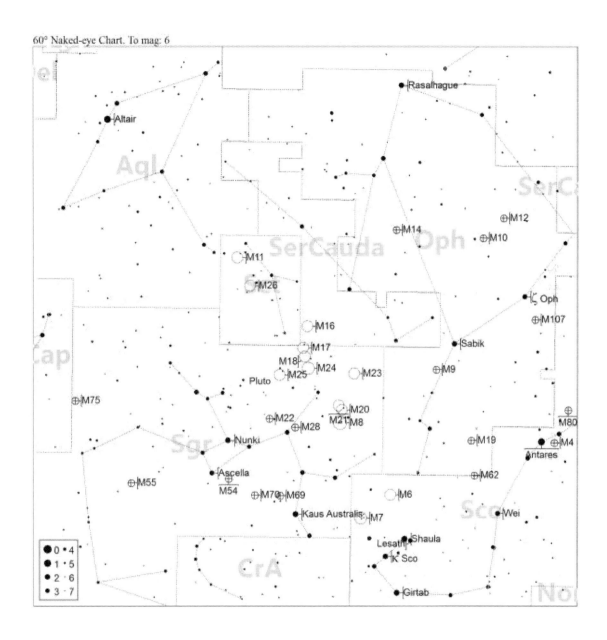

Messier 17: The Omega Nebula

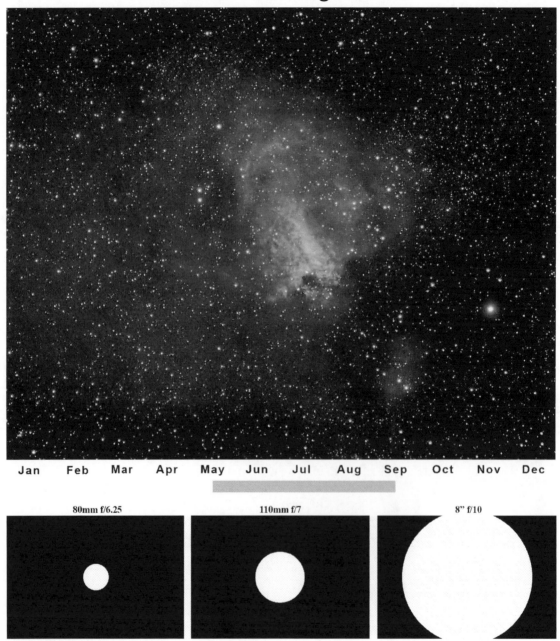

About the target: This ionized hydrogen gas cloud region in Sagittarius was discovered in 1745 by Swiss astronomer Philippe Loys de Chéseaux. This is an emission nebula meaning that it "glows". This glow is believed to be generated from approximately 35 stars hidden behind the gas. Sitting some 5,000 or more light years away and spanning 15 light years across, this is a very easy nebula to image or view in a small scope or even binoculars.

Where it is:
Right ascension: 18h 20m 26s
Declination: −16° 10′ 36″

Imaging the target: This target is pretty much a duplicate of M16 as far as shooting and processing is concerned. There is not as much concern for detail because there is no world famous area like the "pillars of creation" in M16, although the detail level here is probably just as nice. One serious difference is that this target seems to have a few brighter stars that are more easily blown out so care when stretching and exposing is a little more important. I lowered my exposure times here from the 480 seconds I used on M16 to 300 seconds here, both at ISO 800.

I know you are tired of hearing this but I am saying this again; if you are shooting color here, this nebula is much like M16 and M17 in that it is not bright red as many people seem to think. In regards to M17 for example, according to the *Students for the Exploration and Development of Space*, "The color of the Omega Nebula is reddish, with some graduation to pink. This color comes from the hot hydrogen gas which is excited to shine by the hottest stars which have just formed within the nebula. However, the brightest region is actually of white color, not overexposed as one might think." So don't spend a lot of time trying to make it redder than it appears, unless of course you just want to.

I put a little more time and effort into this target than many people, so much so it is the cover of this book, mainly because most people don't. If you like being the odd man out, this is a great target to play with.

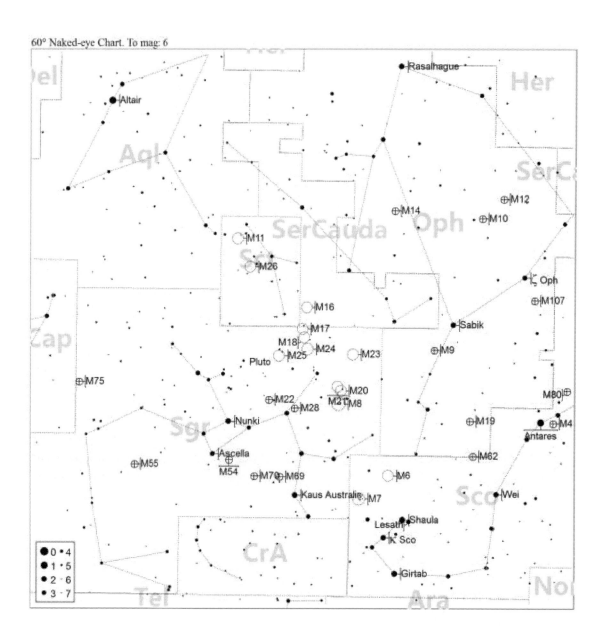

Messier 18:

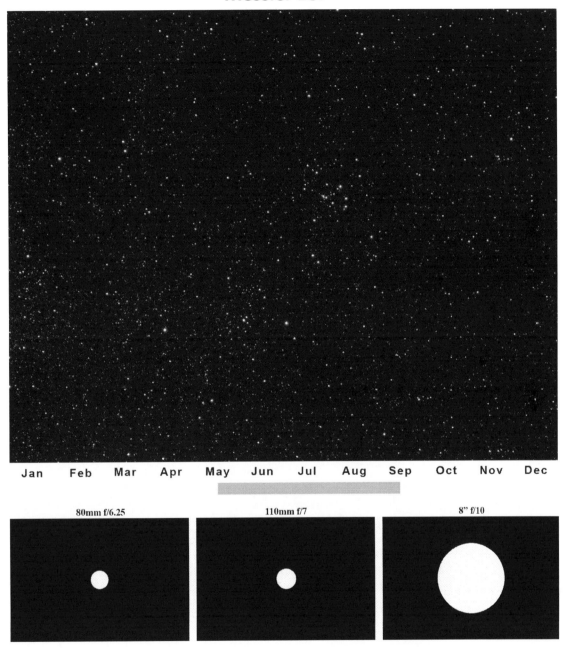

About the target: In the constellation of Sagittarius sits this open cluster of stars almost 5,000 light years away and 18 light years in diameter. Discovered by Charles Messier on the 3rd of June, 1764, the twenty or so stars that make up this cluster are considered young at only about 32 million years old. The cluster looks a little boring and this is not helped by the fact that it is situated between M24 and M17, both awe inspiring targets.

Where it is:
Right ascension: 18h 19.9m
Declination: −17° 08′

Imaging the target: This is one of the objects in the Messier list that is just boring, sorry. The open cluster is not very remarkable in any respect, either visually, or imaged. That being said, your best bet is to expose for the deep end and hold off on the stretching to preserve your star colors if you are shooting color. This will help develop for the only redeeming quality here, a reasonably rich and colorful background of stars. On my setup I would go for a lot of frames of about 240 seconds at ISO 800 or so, stack these and then stretch to maximize star colors by stretching the gray point and leave the white and black points about where they are.

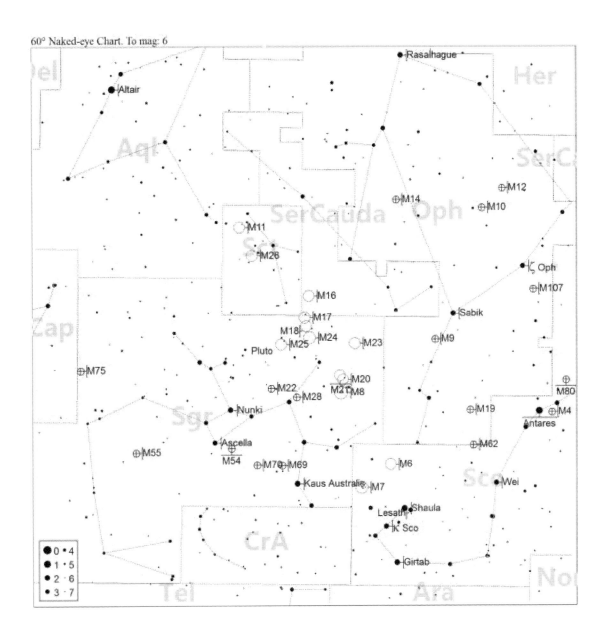

Messier 19:

About the target: On the 5th of June, 1764, Charles Messier discovered this globular cluster in Ophiuchus approximately 29,000 light years away and spanning some 140 light years across. Probably due to dust clouds obscuring part of the light emitted from this cluster, much like M9, this globular appears oblate or squashed from top to bottom as if someone sat on it. Looking around this area one can find several more globular clusters not in the Messier list within 1.5 degrees or so in multiple directions.

Where it is:
Right ascension: 17h 02m 37.69s
Declination: −26° 16′ 04.6″

Imaging the target: Here is a globular cluster which presents a nice background of stars and a pretty nice spread. It is not as well defined in the center as I would like and I think that is more due to my large field of view with my equipment than anything else. In something with a focal length above 1200mm or so this would probably be a really nice target, possibly benefitting from two or three different sets of exposures layered together to catch both the core and spread.

The background of stars is pretty colorful with lots of oranges, yellows and reds so be careful with your stretching as they could easily all blow out to white.

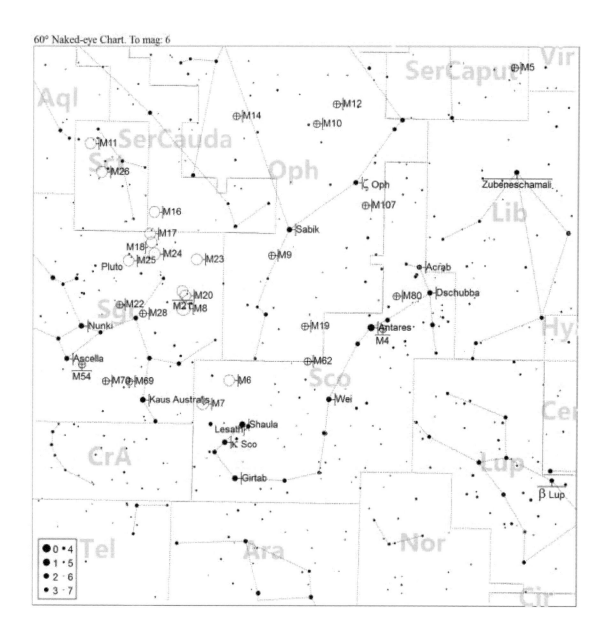

Messier 20: The Trifid Nebula

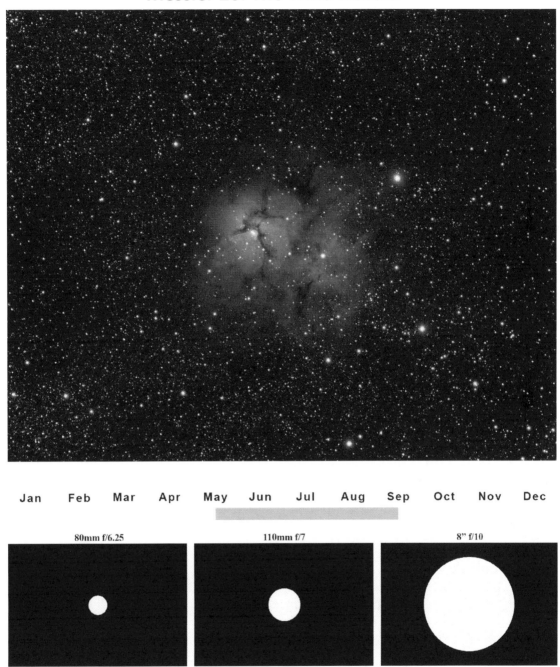

About the target: This fantastic object located in the constellation Sagittarius is actually an open cluster of stars, an emission nebula (red nebula), a reflection nebula (blue nebula) and a dark nebula (the gaps in the nebulas) all in one place. Messier discovered this object on the 5th of June, 1764. In 2005 NASA discovered just under 200 baby stars in different stages of development using the Spitzer Space Telescope. While a fairly small object sitting 5,200 light-years away, it is still visible with small telescopes and is usually colorful even without imaging in a medium sized scope.

Where it is:
Right ascension: 18h 02m 23s
Declination: −23° 01' 48"

Imaging the target: The Trifid Nebula is actually a little harder than it may seem as the nebula portion, especially the blue portion, is rather tricky to stretch. The problem is that the red portion of the nebula is brighter than the blue, so if you shoot/process for the red side, the blue is very difficult to see. If you shoot/process for the blue side, the red side blows out. Either way there are some bright stars right in close that are hard to keep color. Add to all of this the fact that it is starting to warm up a little when M20 is good to shoot and you can have issues with a moderate amount of thermal noise as well.

In this image I attempted to solve the situation by shooting at a higher ISO (1600 in this case) for shorter exposures (150 seconds in this case) and then stretch the midpoints pretty good using a high number of lights (30 in this case) to help take care of any extra thermal or high ISO noise. I didn't completely fail but the blue did not process as well as I had hoped, probably due to the high ISO also causing a lower dynamic range which severely hampered my ability to pull the dark blues out of the black background.

Overall the approach worked more than it didn't so that may give you an interesting place to start.

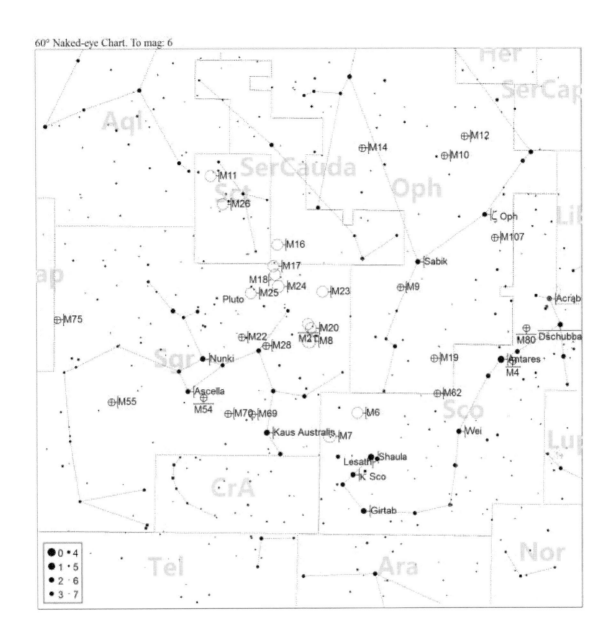

Messier 21:

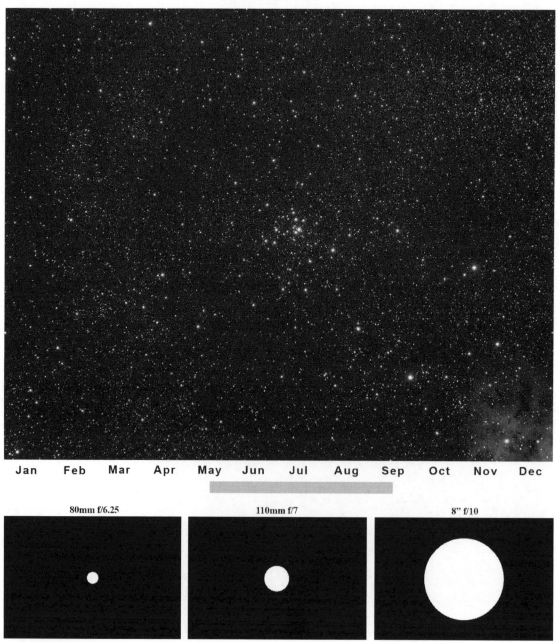

About the target: This open cluster in Sagittarius is considered very young at only 4.6 million years old. Discovered by Charles Messier on the 5th of June, 1764, it is just over 4,200 light years away and contains somewhere near 50 stars. Its close proximity to M20 allows both to fit into the field of view of many amateur telescopes.

Where it is:
Right ascension: 18h 04.6m
Declination: −22° 30′

Imaging the target: Right next to M20 is the open cluster M21. While not totally unremarkable, it is not overly interesting either. When grouped with M20 it can be a very interesting addition to the image which really does not need any special attention when processing. When shot on its own you can go a little deeper with longer exposures if you like, but be careful about blowing out the stars in the bottom right side as they are brighter than the stars in the center of the cluster. While processing in color you can get some interesting red/orange stars in the field which offsets the primarily blue members of the cluster well.

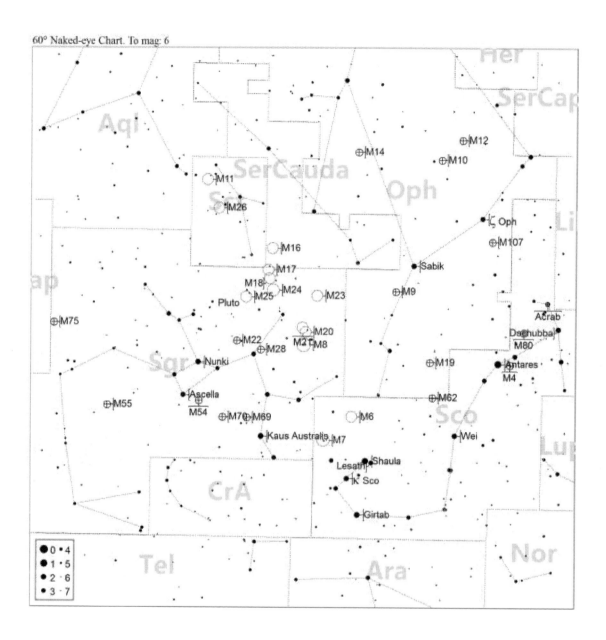

Messier 22: The Sagittarius Cluster

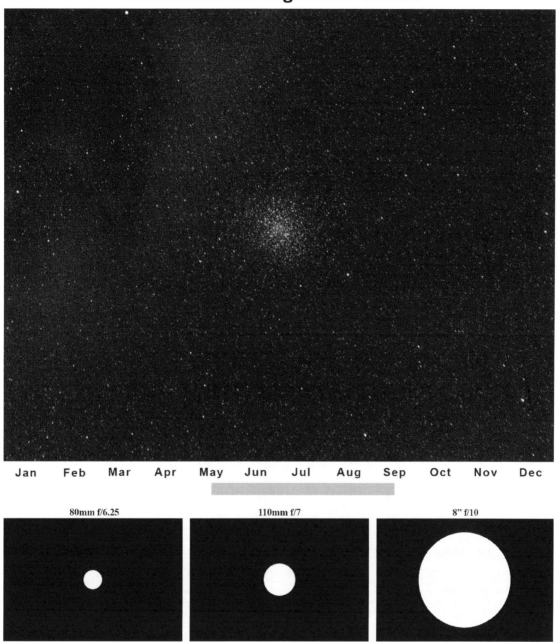

About the target: German astronomer Abraham Ihle discovered M22 in the constellation of Sagittarius, the first globular cluster ever discovered, in 1665. It is approximately 10,000 light-years from Earth and is roughly 100 light-years across. One of the brightest globular clusters in the night sky, Messier noticed and cataloged it on the 5th of June, 1764. According to studies by American astronomer Harlow Shapley in the 1930s, this cluster is home to some 70,000 stars.

Where it is:
Right ascension: 18h 36m 23.94s
Declination: −23° 54′ 17.1″

Imaging the target: This is one of the loosest globular clusters in the Messier list so you really do not have to worry about blowing out the core very much. It is also pretty bright, so shorter exposures can be used although at this time of year it is starting to cool down a little so that isn't much of a problem. One advantage of shooting this target with shorter exposures is that since it is pretty bright and is on a rich star filled background, it is fairly easy to artificially increase the contrast between the cluster and the background.

Longer exposures show more stars and make for a really nice background, but then the cluster can kind of get lost in the rest of the field. Try it both ways and see which you prefer.

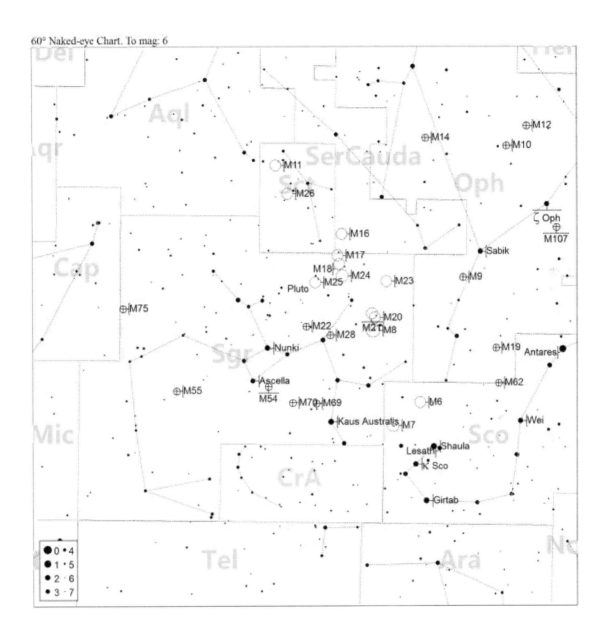

Messier 23:

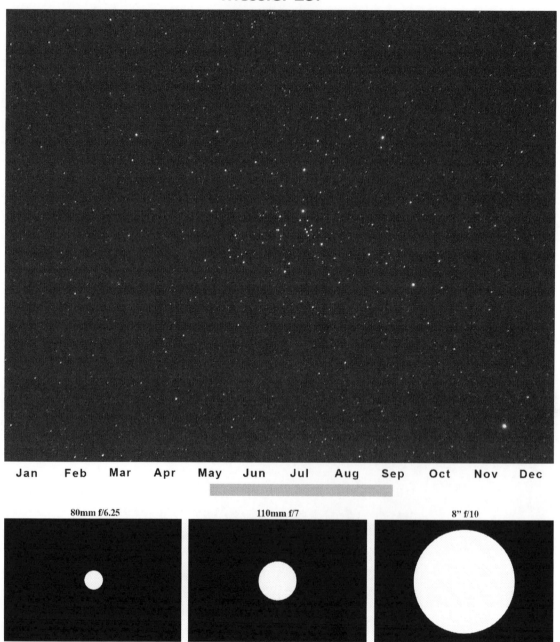

About the target: Only 2,150 light years away in the constellation Sagittarius sits this little open cluster of about 150 stars in an area around 15 light years across. Discovered on the 20th of June, 1764 by Messier, who noted "A star cluster, between the end of the bow of Sagittarius & the right foot of *Ophiuchus*, very near to 65 *Ophiuchi*, according to Flamsteed. The stars of this cluster are very close to one another."

Where it is:
Right ascension: 17h 56.8m
Declination: −19° 01'

Imaging the target: Unfortunately this is another one of the more boring objects in the Messier catalog and I have shot it only one time because of that. The biggest trick to shooting this cluster is to get the actual cluster separated from the background of stars. Too much exposure and/or stretching and it looks like one big field of stars. Too little and there is nothing there. I would probably suggest a medium length exposure at around 240 seconds for my equipment at ISO 800 with very minimal processing.

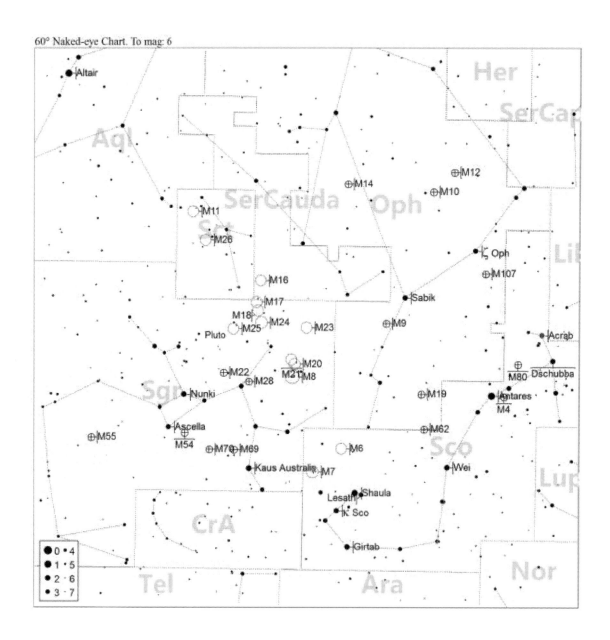

Messier 24: The Sagittarius Star Cloud

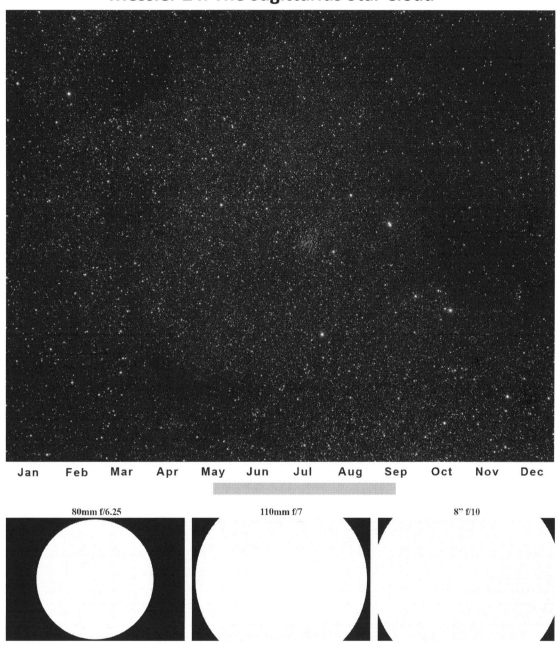

About the target: With nothing more than a simple pair of binoculars, one can see up to 1,000 stars at one time in their field of view looking at the Sagittarius Star Cloud. Discovered by Charles Messier on the 20th of June, 1764, this is actually a huge cloud of stars some 10,000 light years away and spanning over 700 light years in length which is a portion of the Sagittarius arm of the Milky Way galaxy.

Where it is:
Right ascension: 18h 17m
Declination: −18° 29′

Imaging the target: Right on the heels of the boring M23 comes something very exciting. With most clusters there is a central region you concentrate on, and the background field might get a little attention if it is nice. Here we have the opposite where the subject is the background, and what a background it is. Full of blue, orange, red and yellow stars you can just get lost looking at this target in the eyepiece or in a picture. Shoot a midrange exposure, in my case 240 seconds at ISO 800, and shoot quite a few frames. You do not want noise to damage any of the faint stars or stretching to remove any star colors. This one needs to be as right in the camera before processing as you can get it and it also needs pretty good polar alignment and tracking to keep the stars as small as possible.

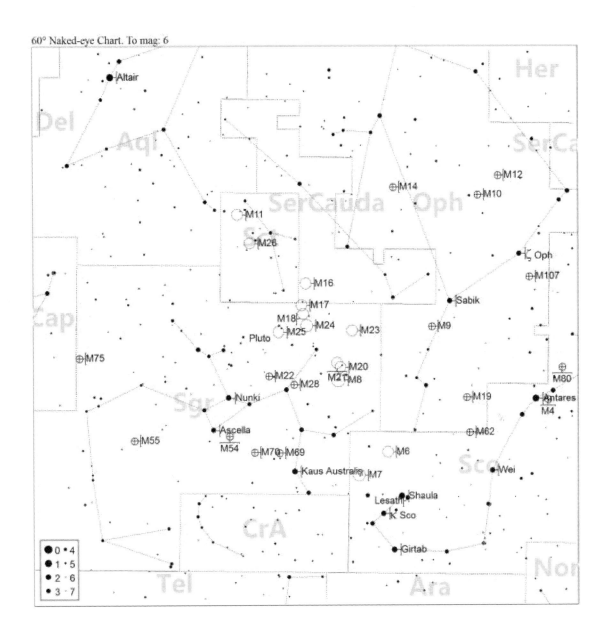

Messier 25:

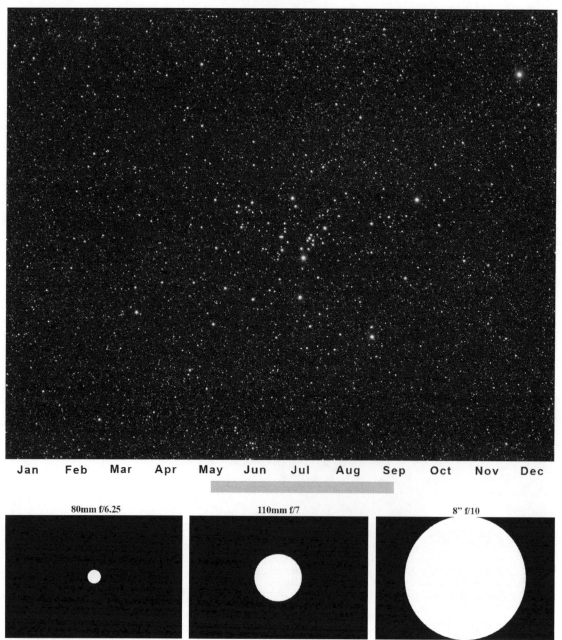

About the target: Discovered as early as 1745 by Philippe Loys de Chéseaux this open cluster in Sagittarius spans about 20 light years across and is approximately 2,000 light years away. This cluster is easy to view in binoculars and has been extensively studied and cataloged by people such as Messier in 1764, Bode in 1777, Herschel in 1783, Smyth in 1836, Webb in 1859, Schmidt in 1866, Dreyer in 1908, and Bailey in 1908, among others.

Where it is:
Right ascension: 18h 31.6m
Declination: −19° 15′

Imaging the target: Another semi-boring open cluster. It has a reasonable concentration in the center and a reasonable background of stars. I would suggest something like a medium exposure, about 240 seconds for me, and a light touch on processing to preserve as much star color as possible, especially on the brightest stars which vary in colors substantially. Be careful of that one star in the upper right, it can blow out really easy and may need to be cropped if you are going to try to go deep and get a serious background of stars.

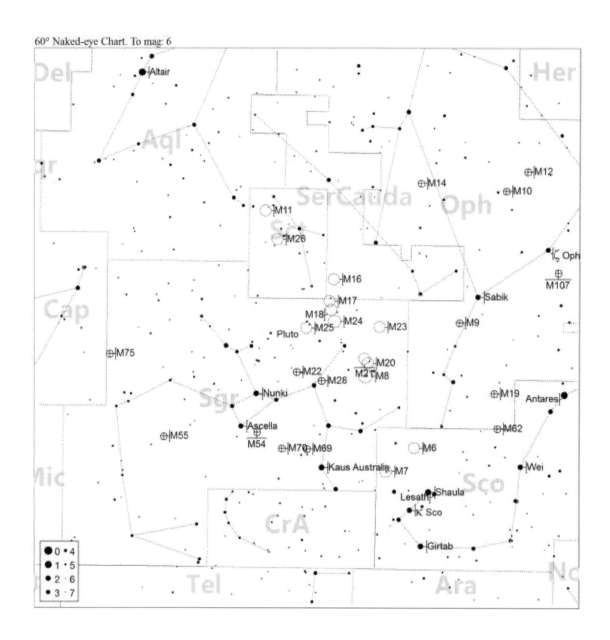

Messier 26:

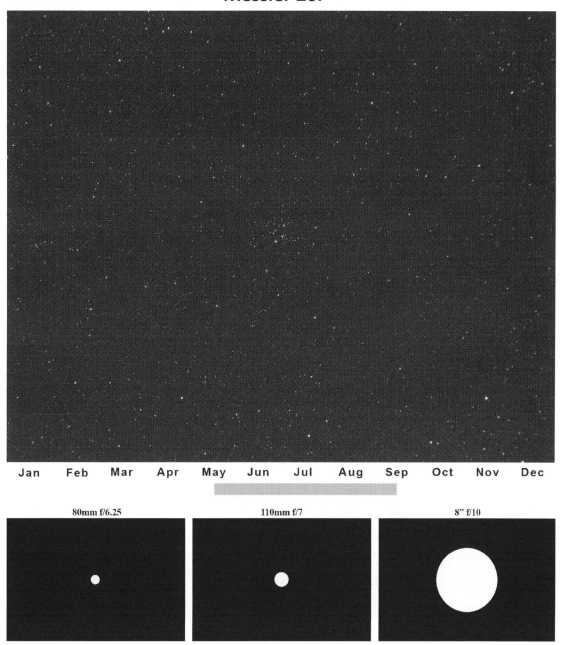

About the target: This open cluster is in the constellation Scutum and is some 5,000 light years away and spans 22 light years across. Note the rather dark section of the inner cluster to the lower right, this is most likely due to some dust cloud between us and the cluster obscuring our vision of the stars that are most likely there. M26 was discovered by Messier on the 20th of June, 1764.

Where it is:
Right ascension: 18h 45.2m
Declination: −09° 24′

Imaging the target: Here is an open cluster I need to go back and reshoot as the image I have here does not do it justice. Many of the open clusters in the Messier list are pretty boring. This one can be really interesting. With a little deeper exposure of around 240 seconds for my equipment at ISO 800, it should be possible to really bring out the patterns and colors of this target. Its best quality is if you are shooting color, it has a wonderful array of blues, oranges and white stars all mixed in together right in the central area of the cluster, as well as spread throughout the background field. Pay special attention to preserving these colors and you should be able to make a really nice image. If you are shooting monochrome on the other hand, this will unfortunately be another boring target.

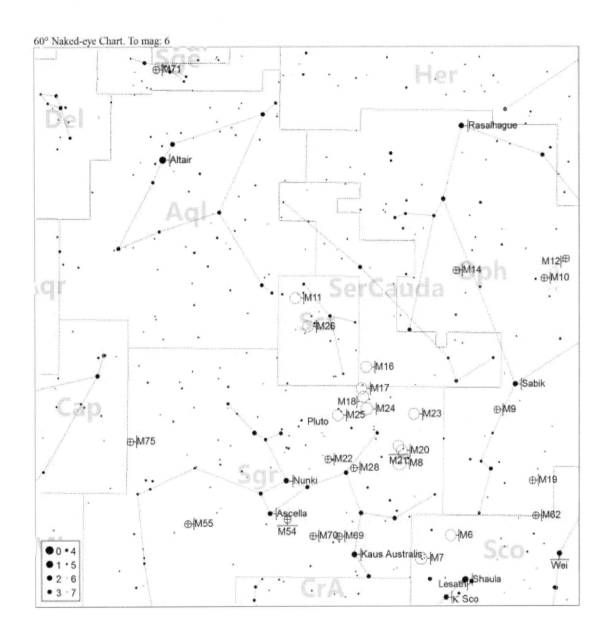

Messier 27: The Dumbbell Nebula

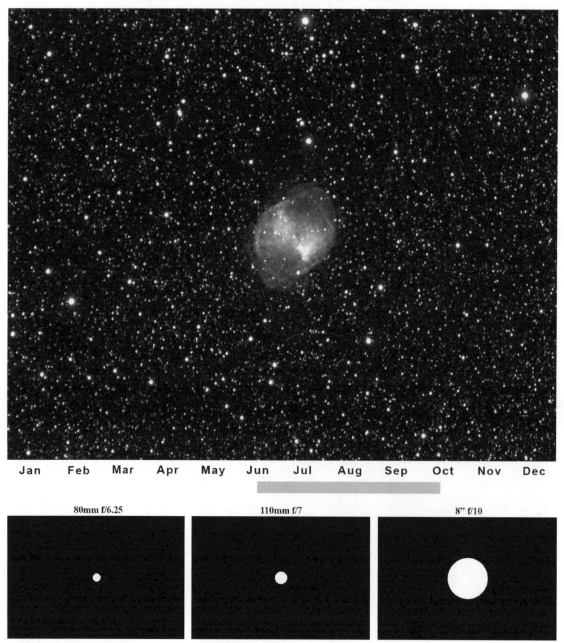

About the target: Messier discovered the first planetary nebula, M27, on the 12th of July, 1764 and noted "Nebula without star, discovered in Vulpecula … it appears of oval shape, & it contains no star." This planetary nebula sits in the constellation Vulpecula and is some 1,400 light years away with approximately a 3 light year diameter. The ionized gas shell you see was most likely ejected from a late term star.

Where it is:
Right ascension: 19h 59m 36.340s
Declination: +22° 43' 16.09"

Imaging the target: The Dumbbell Nebula is a pretty easy target to capture, but then you will want to get some real detail in there and it becomes a little trickier. The first problem you run into here is that the best time of the year to shoot it is also the hottest. This means you will have a serious issue with increased thermal noise. To make this problem even worse, to get the details you need, you will have to go deep; 480 second exposures at ISO 800 for me. That makes the thermal noise exponentially worse. Now let's heap gasoline on the fire and make the target fairly small and details even smaller so we have to have our tracking and alignment dead on.

The good news is that if you want a bluish blob in the sky picture, this is a very bright easy target. The bad news is that if you want the red on the sides and the red X pattern on the back side of the nebula to show up, be prepared to really work at this target. I love targets that can be shot over and over with different levels of details showing up depending on how I choose to shoot it, and this is a fantastic one for that.

My suggestion is to shoot long exposures (480 seconds for me) at a reasonable ISO (800 for me) and stack as many of them together as you can get (25 or more would be great). Now stretch for maximum detail in the nebula and try not to blow out the stars either by making sure you use curves to control the areas being stretched or by using layers to isolate the nebula from the surrounding stars.

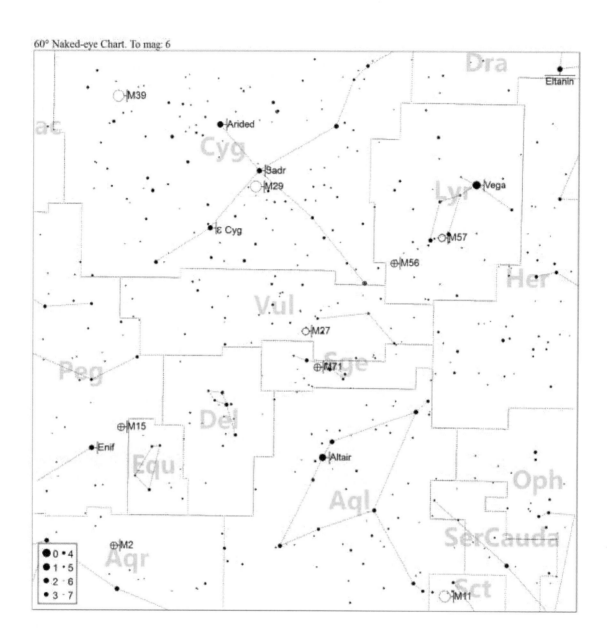

Messier 28:

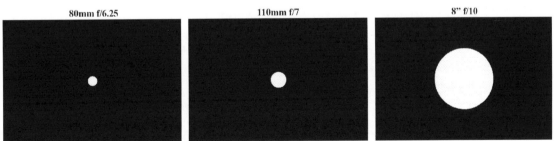

About the target: This small globular cluster in Sagittarius is set on a rich star filled background and can be seen through a reasonable pair of binoculars. Almost 18,000 light years away and spanning 60 light years across, this globular was discovered by Charles Messier on the 27th of July, 1764. M28 is noted as being the location for the discovery of the first millisecond pulsar inside a globular cluster in 1986. Since then, there have been a total of 12 discovered in this cluster, making it the third most dense population of millisecond pulsars known.

Where it is:
Right ascension: 18h 24m 32.89s
Declination: −24° 52′ 11.4″

Imaging the target: Normally I do not like small nondescript globular clusters, especially when the core is only semi-defined. This one however is an exception because of the beautiful background field of stars. While most of the stars in the cluster are yellow and white, you can see a nice selection of blue and orange stars in the rest of the image. It even looks like there could be several open clusters in the field as well. Although the globular is unremarkable, when you add in the rest of the field of view, you wind up with a really nice image whose sum is greater than any of its parts.

Now to shoot this little guy you need to pay particular attention to the stars and their color by shooting short exposures and lots of them. Then remember to stay away from the white point while stretching the image. I went for ISO 800 and 150 second images and it worked well for me. You might also want to make sure you have a really good polar alignment and tracking here so that the stars are as small as possible as it really helps this type of image.

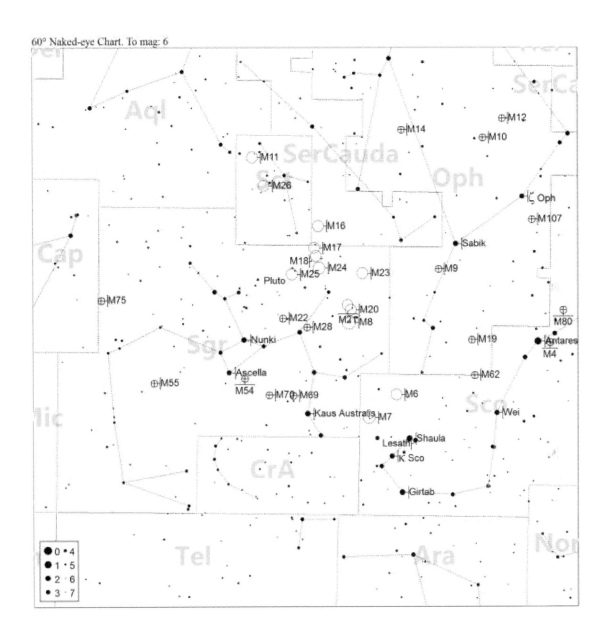

Messier 29:

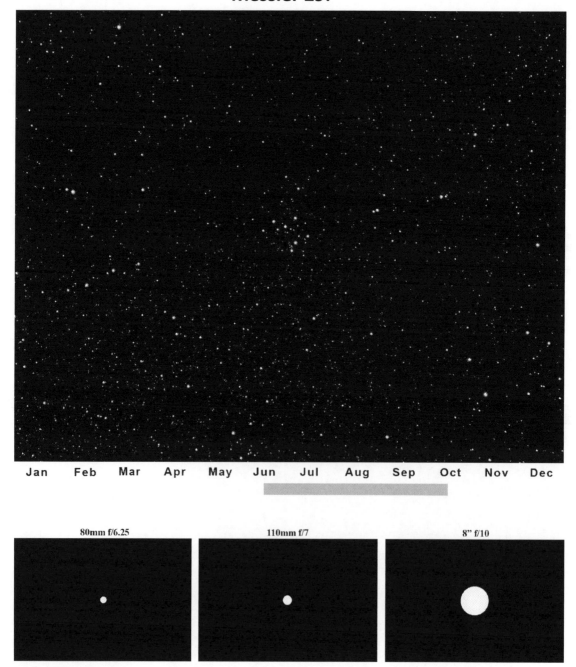

About the target: This 10 million year old cluster was discovered by Messier on the 29th of July, 1764 and can be seen using a very small telescope or binoculars. The cluster is believed to be some 4,000 light years away. Messier's notes state "A cluster of 7 or 8 very small stars" which is an apt description.

Where it is:
Right ascension: 20h 23m 56s
Declination: +38° 31.4'

Imaging the target: I have tried and tried to make this target interesting, and I am not ashamed to say I have failed miserably every time. This image is 14 240 second ISO 800 images stretched too far to try and bring out more stars but it just didn't happen. In fact, I have never seen a really interesting shot of this target, although I have seen many better than mine.

I might suggest shooting a lot more frames, maybe 30 or more, and keeping the exposure time at medium or even lower. It doesn't help that this target is up in the hotter part of the year so thermal noise is going to try to kill a lot of background stars. Whatever you do, keep any stretching to a minimum as these stars seem to lose their color rather rapidly, probably another effect of the thermal noise.

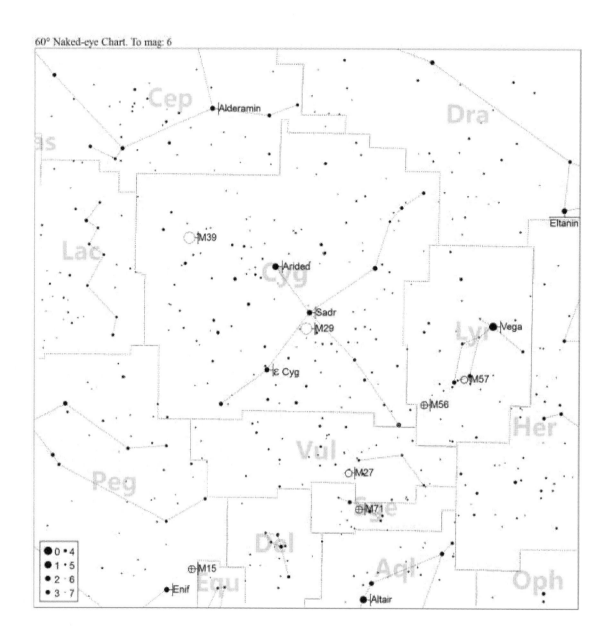

Messier 30:

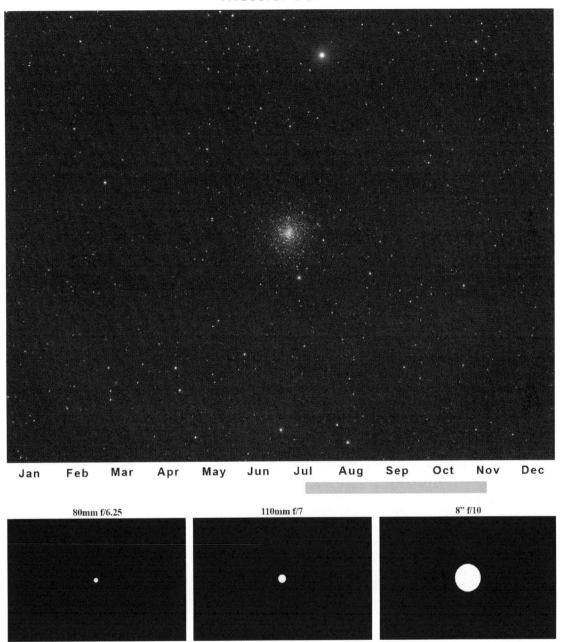

About the target: Roughly 29,000 light years away in the constellation of Capricornus sits this globular cluster which spreads out over 90 light years in diameter. In color images you will note how blue the center of the cluster is. This is caused by the extreme density of stars which can result in a process called mass segregation, where the central region contains a greater amount of higher mass stars, hence they are blue. Messier discovered this little gem on the 3rd of August, 1764.

Where it is:
Right ascension: 21h 40m 22.12s
Declination: −23° 10′ 47.5″

Imaging the target: If you want a fairly detailed image of a globular cluster that is clearly separated from the background, this is your target. Without several sets of exposures, this target shows a detailed core and moderate spread with very little around it except that big honking bright star right above it (magnitude 10.25 HIP 107128). Not only that, but it will respond to fast exposures very well (120 seconds or shorter at ISO 800 for me).

Now if you want to do better, this globular actually has a pretty nice spread which can be easily captured with 240 second images on my equipment. This unfortunately also tends to blow out the core so you have to use HDR or multiple layers for this to work. Even then, the background is fairly black giving you excellent contrast between target and background.

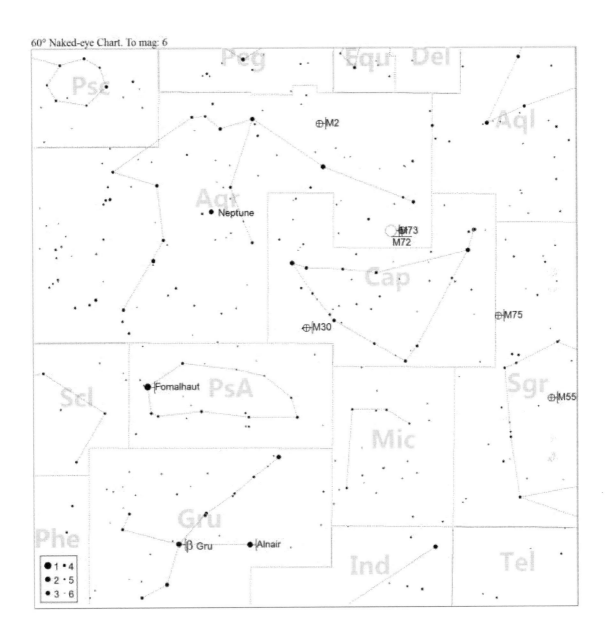

Messier 31: The Andromeda Galaxy

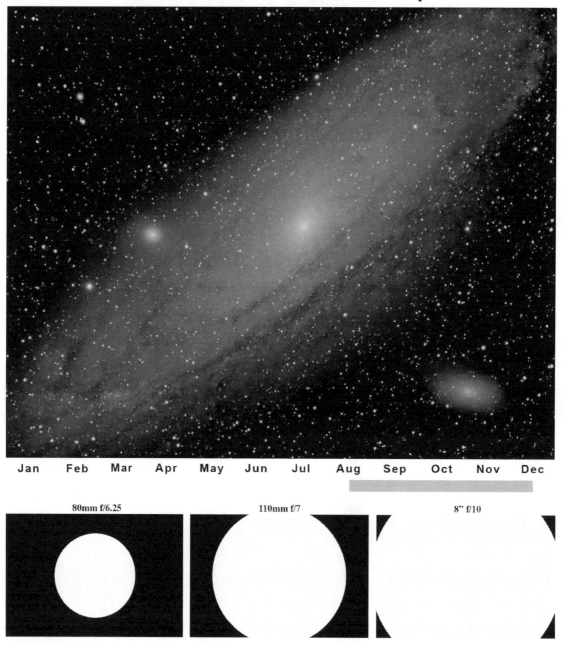

About the target: The Andromeda Galaxy is the closest spiral galaxy to our own at 2.5 million light-years from Earth and sits in the constellation … wait for it … Andromeda. Part of the "local group" of galaxies including the Milky Way, Triangulum Galaxy, Sextans A, Leo A and many others it is the largest. In approximately 4.5 billion years it will collide with the Milky Way in what should be an astounding event. This is the only galaxy other than our own which is viewable without a telescope in a dark sky. It has a recorded history going back to the first century AD where it was noted by Persian astronomer Abd al-Rhman al-Sufi as a small cloud.

Where it is:
Right ascension: 00h 42m 44.3s
Declination: +41° 16′ 9″

Imaging the target: There is no big secret to imaging this bright object; it is almost all in the processing. Of course you have to take images you can process first so let's talk about that for a minute.

This is a huge bright target and your images should be as deep as you can get them without completely blowing out the central core region. Then, you need to shoot a lot of them, at a bare minimum, 20 frames, preferably closer to 50. You also need to shoot a low ISO so that you have the dynamic range to stretch out the faint outer dust lanes including the blue halo around the outside. Lastly, you need to shoot this target when it is as cold as you can get it outside to minimize thermal noise.

Once you have all of that, this target is going to require some finesse with the curves stretch and you will need to stretch it a little at a time over a fair number of passes, probably somewhere near 10 stretches or more. Take your time or you will overdo parts of it. Be very careful here, it is extremely easy to overstretch it. I like to do a small stretch, then copy the layer, and then do another small stretch, and so on. This makes sure that when I overdo it (not if, but when) I can easily go back without having to start over. I can also compare stretches by hiding and revealing layers.

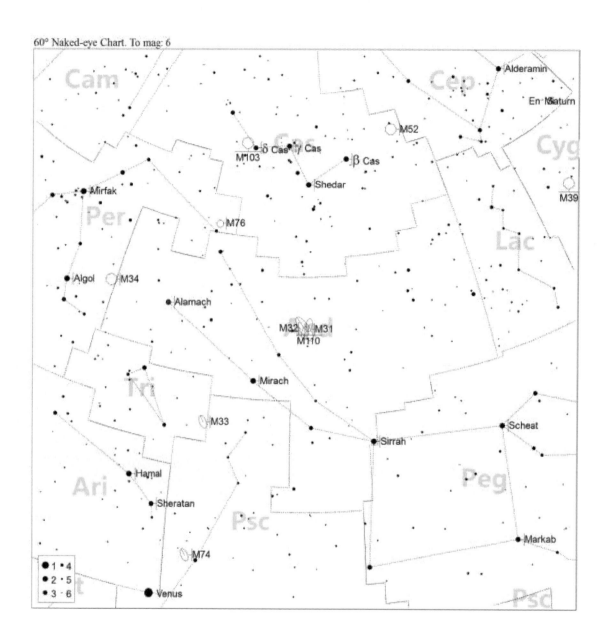

Messier 32:

About the target: If this image looks like a crop of M31, it is! M32 sits just off to the side of M31 but is its own galaxy, just smaller. Originally discovered by French astronomer Guillaume Le Gentil around 1749, it is only 2.65 million light years away but is tiny in comparison to the massive Andromeda galaxy. It is the large fuzzy blob just in the center of the frame that looks like a bloated star.

Where it is:
Right ascension: 00h 42m 41.8s
Declination: +40° 51' 55"

Imaging the target: There are no "fantastic" images of M32, as it is a small satellite galaxy to M31 and you are going to get nothing but a fuzzy blob no matter what you do. Why? Mostly because it is a dwarf elliptical galaxy, which by definition is a small fuzzy blob. The Hubble space telescope took images of M32. Care to guess what they looked like? Yep, fuzzy blobs.

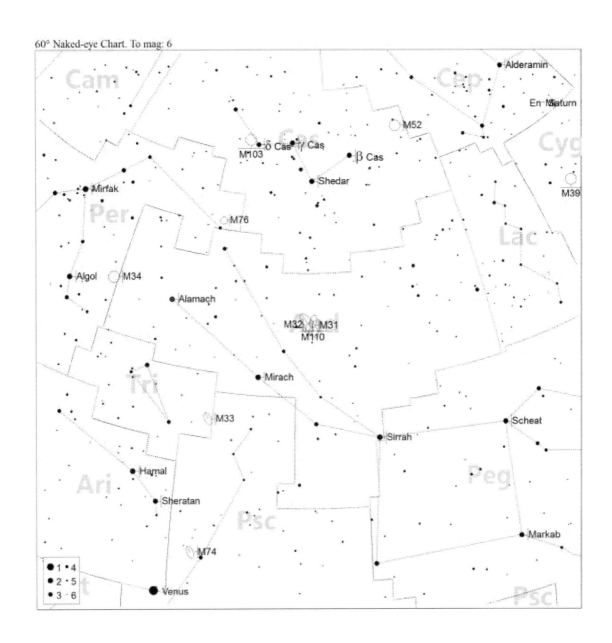

Messier 33: Triangulum Galaxy

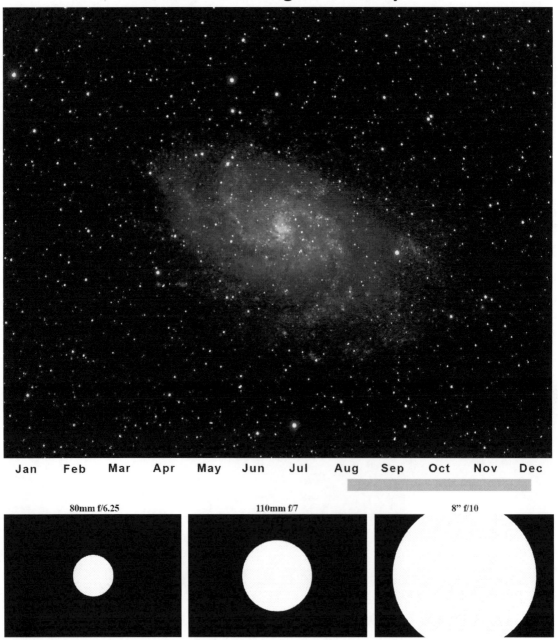

About the target: The Triangulum Galaxy is the third largest member of the "local group" of galaxies which includes the Milky Way, Andromeda Galaxy, Sextans A, Leo A and many others, and sits about 3 million light-years away in the constellation Triangulum. In very dark skies with no light pollution, this can actually be seen with the unaided eye although it is very dim. It was discovered by the Italian astronomer Giovanni Battista Hodierna in the early to mid 1600s.

Where it is:
Right ascension: 01h 33m 50.02s
Declination: +30° 39' 36.7"

Imaging the target: The biggest problem I have with M33 is that when it is in a good place in the sky, here in Texas it is hot. I don't mean hot like "oh, it's hot, I better wear shorts" I mean hot enough to get heat exhaustion at midnight hot. This means even though this is a nice, big and fairly bright galaxy, for me it is very difficult to get a good image from all the thermal noise in the images.

The solution of course is to shoot lots and lots of frames. This galaxy is bright in comparison to many other objects, but you still need to go deep to get the outer arms, and I shoot something like 480 seconds at ISO 800 for that. I would suggest at least 30 or more shots to get something like this with reasonable detail, and the more, the better. Those of you in cooler areas, or areas with lower humidity may fare much better than I have.

Like most large galaxies, you need to be careful stretching the outer arms as you do not want to blow out the center and surrounding stars. I would suggest using more curves than levels for this although an initial stretch with levels may prove useful.

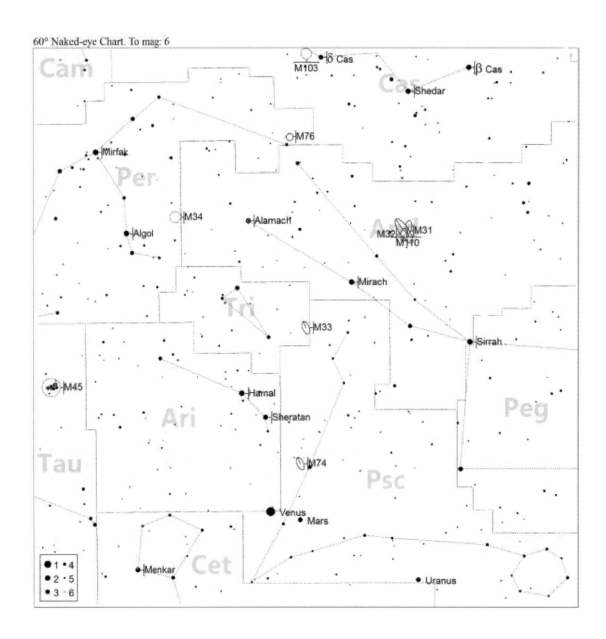

Messier 34:

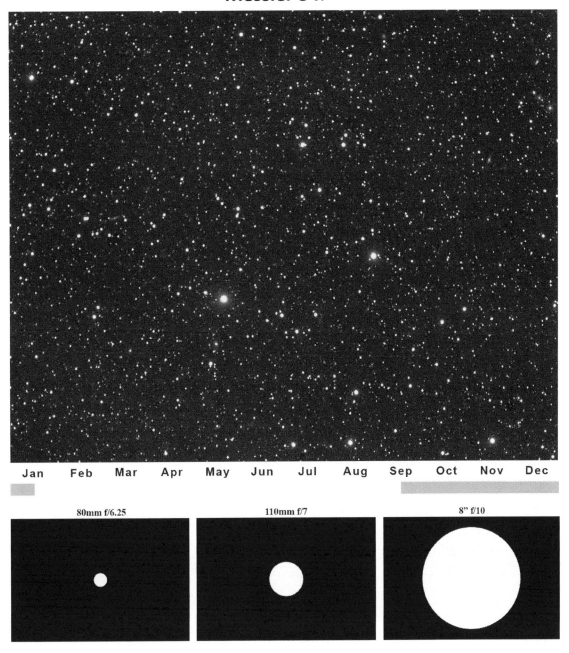

About the target: Most likely M34 was discovered by Italian astronomer Giovanni Batista Hodierna in the early to mid 1600s while Messier cataloged it on the 25th of August, 1764. The cluster is home to some 400 stars, and it resides approximately 1,500 light years away and is about 7 light years across. It is believed to be somewhere near 180 million years old.

Where it is:
Right ascension: 02h 42.1m
Declination: +42° 46′

Imaging the target: This is the epitome of boring open cluster. Don't get me wrong, there are some spectacular open clusters, but Messier was not interested in beauty when he made this list so some of the items are just plain boring. The only thing you can hope for with this is to shoot color and carefully process it to get some nice star colors as there are a few nice yellow/orange giants in the field.

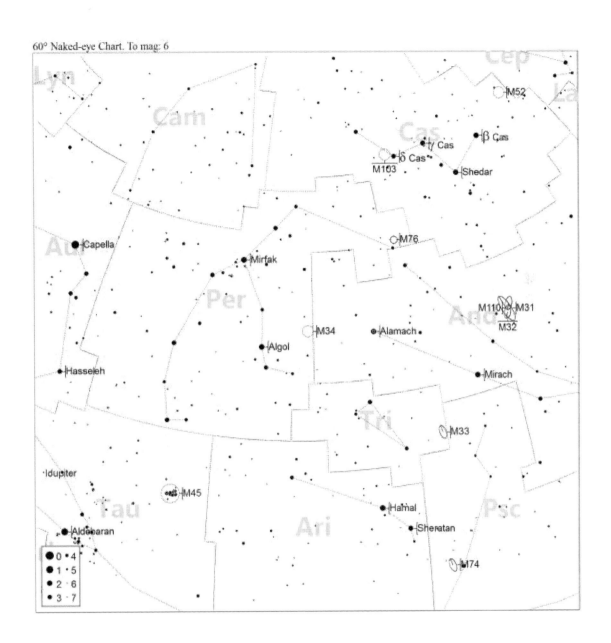

Messier 35:

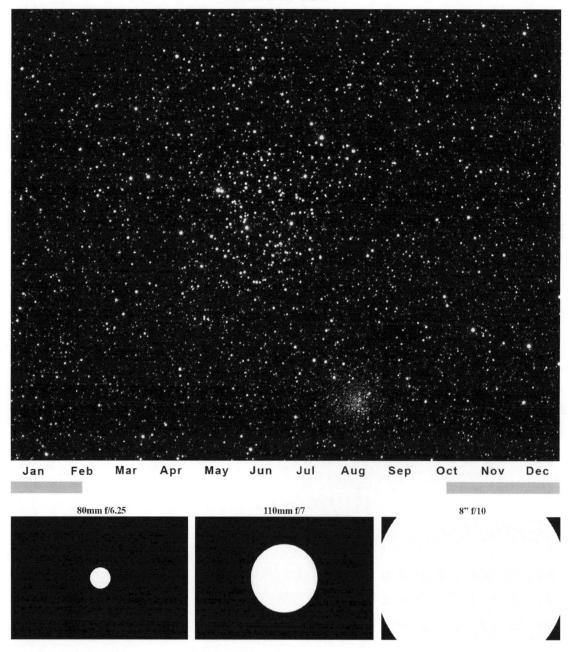

About the target: Some 2,800 light years away in the constellation of Gemini is M35 and it was most likely discovered by Swiss astronomer Jean-Philippe Loys de Chéseaux in 1745. Approximately nineteen years later, Messier cataloged it and made reference that it had already been cataloged by English astronomer John Bevis whose work was completed in 1750. Spanning some 22 light years across, it consists of several hundred stars that are fairly easy to see with the unaided eye from a reasonably dark site.

Where it is:
Right ascension: 06h 09.1m
Declination: +24° 21'

Imaging the target: This is a reasonably interesting open cluster in that it is a little more defined than most, allowing you to see some patterns in the stars. In addition, if you have a slightly wider field of view than I do (maybe with an 80mm refractor or fast Newtonian) and go a little deeper than I did with a longer exposure, you can get NGC2158 in the shot as well which is a nice open cluster that is almost a globular so it provides nice contrast. I recommend starting at somewhere around 240 seconds of exposure at ISO 800 for my setup and adjust from there.

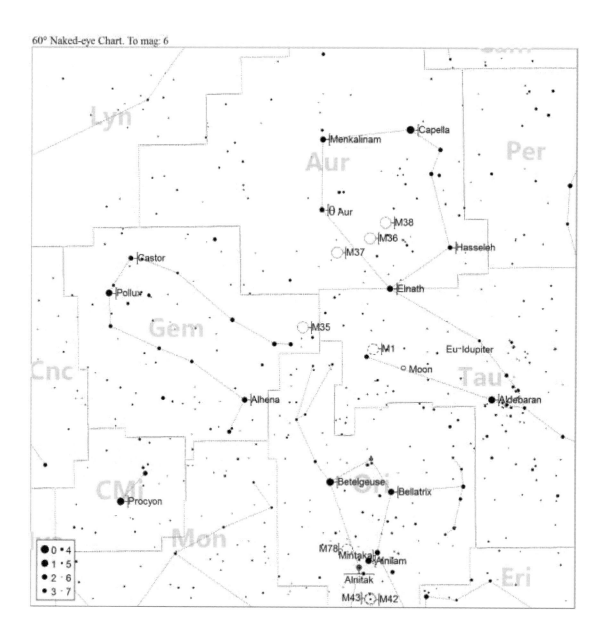

Messier 36:

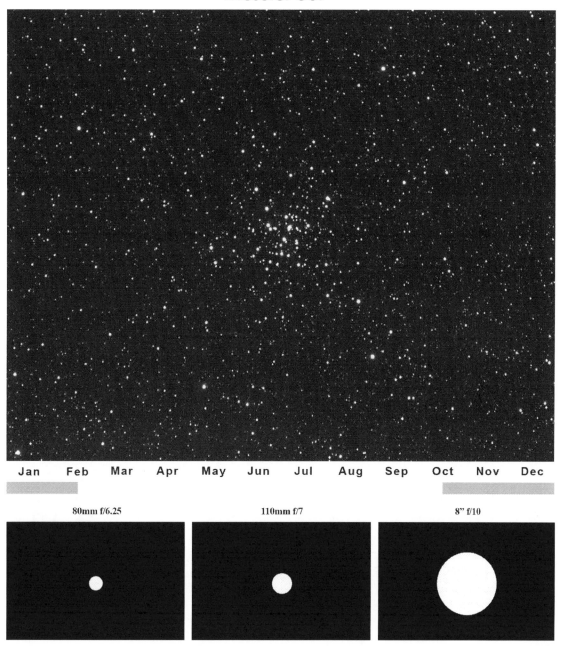

About the target: Discovered by Italian astronomer Giovanni Batista Hodierna somewhere in the early to mid 1600s M36 sits some 4,100 light years away in the constellation of Auriga and is approximately 14 light years across. This little cluster is similar to M45 (without the nebulosity and further away) and contains at least 60 stars. Messier cataloged this cluster on the 2nd of September, 1764.

Where it is:
Right ascension: 5h 36m 12s
Declination: +34° 08′ 4″

Imaging the target: Here is another reasonable example of a pretty well defined open cluster. This is a rather early image from me so it is not very good but even so, you can clearly see the concentration to the center with a nice background field. I shot at 240 seconds at ISO 800 and then overstretched it a little so the stars are a bit blown out. Lightening up on the stretching should give far better results.

With a lot of medium exposures you can get some fantastic colors here are there are a lot of yellow/orange stars dotting the field of mostly blue stars. In addition, if you have a fairly wide field of view you can capture Barnard 226, a nice red nebula off to the side about 1.5 degrees.

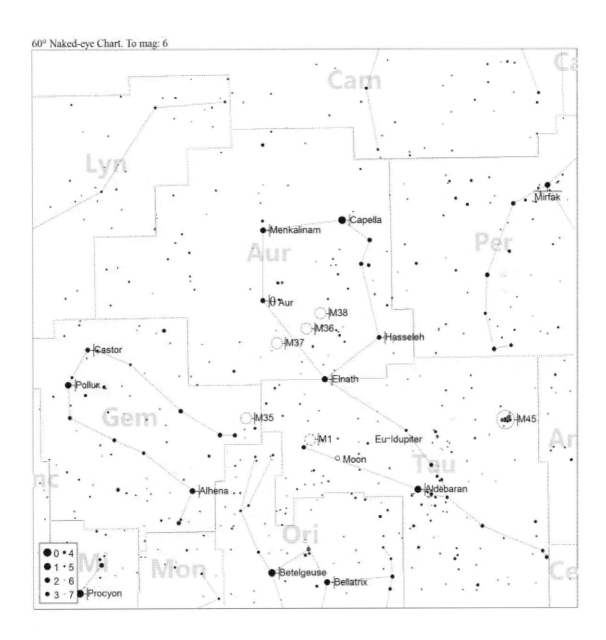

Messier 37:

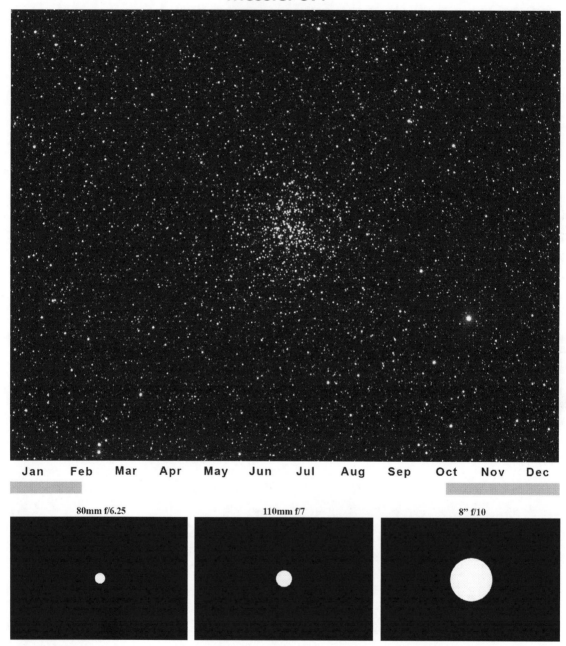

| Jan | Feb | Mar | Apr | May | Jun | Jul | Aug | Sep | Oct | Nov | Dec |

80mm f/6.25   110mm f/7   8" f/10

About the target: Discovered by Italian astronomer Giovanni Battista Hodierna before 1654, it is the brightest open cluster in its area of space. With some 500 stars around 300 million years old it is a wonderful sight to see. This is one of three bright clusters in the constellation of Auriga and is approximately 25 million years old.

Where it is:
Right ascension: 5h 52m 18s
Declination: +32° 33' 02"

Imaging the target: Yet another example of a nice defined open cluster, almost resembling a globular in a wide field shot. Lots of variation of star types makes this a very interesting field. I would suggest a medium length exposure, about 240 seconds for my equipment at ISO 800. Limited if any stretching at all, and then only to the gray point to bring out some of the dimmer stars.

Be careful both in taking the image and in processing of that big bright orange star on the right side as it can blow out really easily.

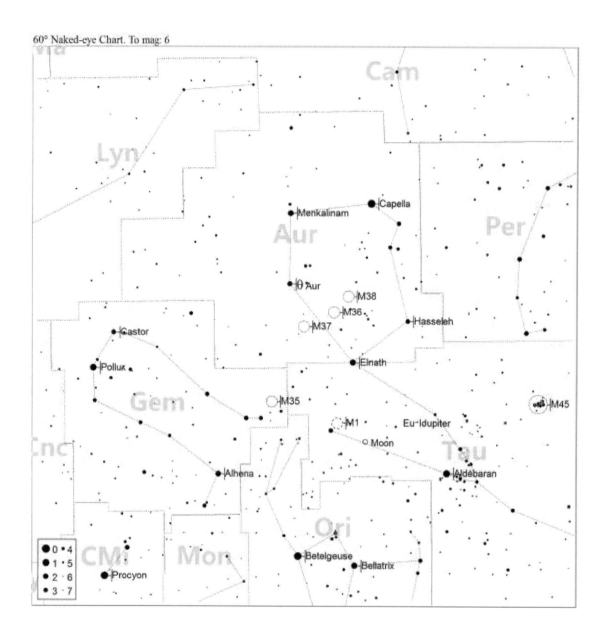

Messier 38:

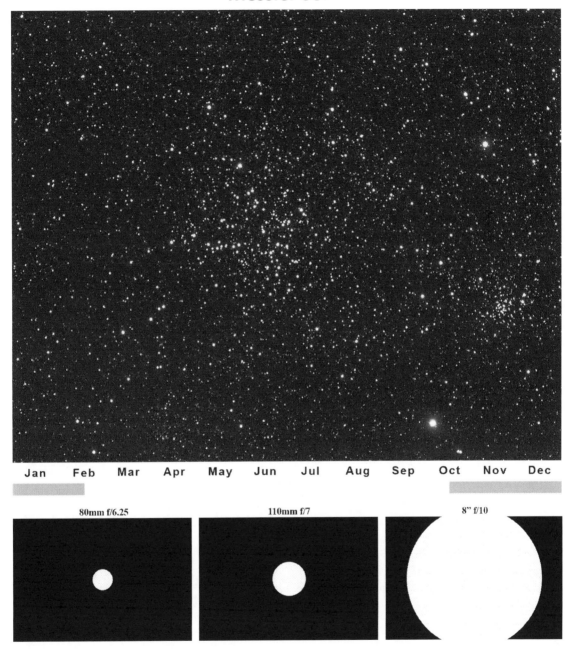

Jan   Feb   Mar   Apr   May   Jun   Jul   Aug   Sep   Oct   Nov   Dec

80mm f/6.25   110mm f/7   8" f/10

About the target: Roughly 4,200 light years away and 25 light years across, this open cluster was discovered by Italian astronomer Giovanni Batista Hodierna in the mid to early 1600s. One of three bright clusters in the constellation Auriga and is believed to be about 220 million years old. It was added to Messier's catalog on the 25th of September, 1764.

Where it is:
Right ascension: 5h 28m 42s
Declination: +35° 51′ 18″

Imaging the target: This is a very interesting area that lends itself to a lot of different approaches. In the first, you could either use a long focal length scope or cropping to get just the central region of the open cluster and this would provide an interesting shot as there appears to be a lot of nice patterns to explore as well as nice coloration.

Next, you could back up a little and include what looks like a second open cluster, and in fact is Herschel H39-7. This is the approach I used in the image included here. You still get the colors of the first approach, you just get more points of interest. You do however need to watch out for that really bright orange star in the lower right, HIP 25476, as it will blow out very easily.

Lastly you could go very wide field and include even more, and then use very long exposures (for me, 480 seconds or more at ISO 800) and carefully stretch the image to reveal a lot of nebulosity in the area that is not really apparent. Be warned however, it could take 50 or more images stacked before the nebulosity really shows up, if at all depending on your location and temperature.

60° Naked-eye Chart. To mag: 6

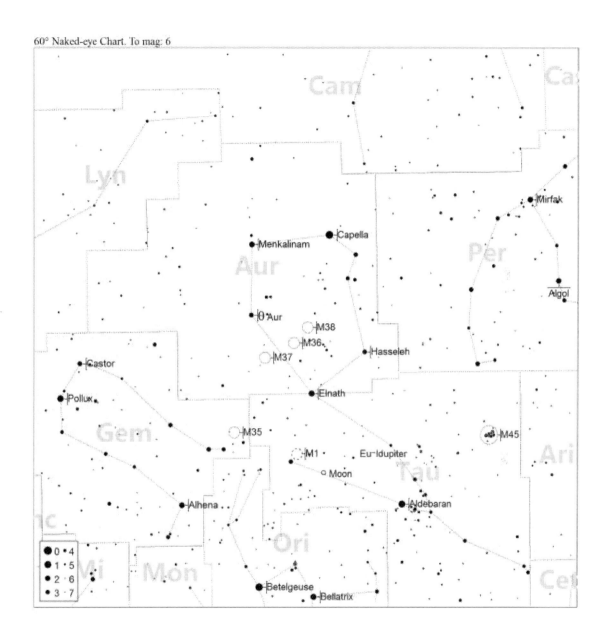

Messier 39:

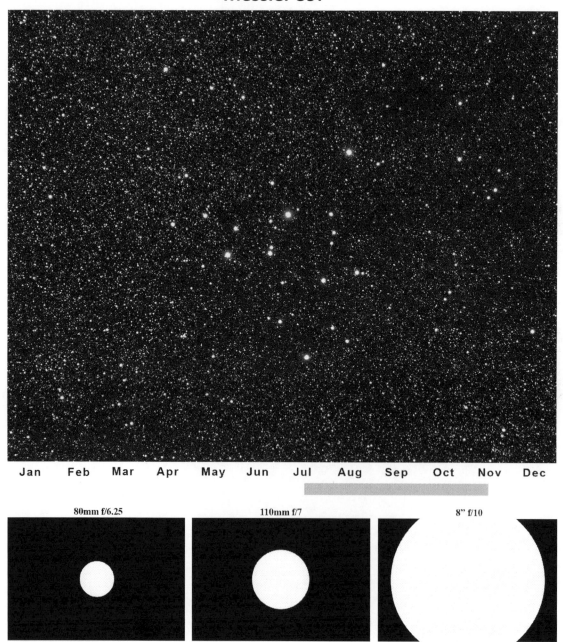

About the target: Messier discovered this cluster in the constellation of Cygnus on the 24th of October, 1764. It is about 800 light years away and it is believed to be between 200 and 300 million years old. Over 30 stars call this cluster home, and they can be fairly easily seen in binoculars and small telescopes.

Where it is:
Right ascension: 21h 31m 42s
Declination: +48° 25' "

Imaging the target: Would you like an open cluster that is both interesting and boring at the same time? Here it is! Boring because of its simplicity; not a lot of stars, no real pattern, not a lot of color variation. Interesting because the background field is quite full without having to shoot deep. A reasonable number of exposures (15 or more) at a medium exposure length (240 seconds at ISO 800 for me) will reveal a background just brimming over with stars and a nice set of brighter stars in the central cluster. Your only real worry is that this is up in a warmer time of year so shoot it late in its cycle to avoid as much thermal noise as possible.

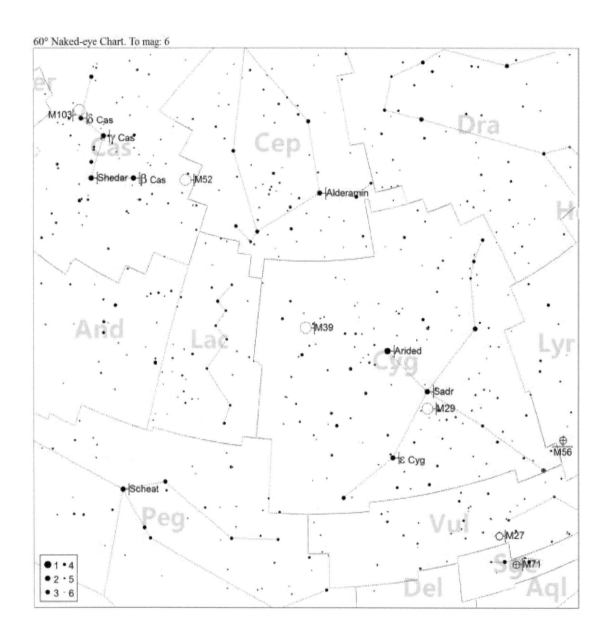

Messier 40: Winnecke 4 Double

About the target: These two stars are separated by approximately 51".7, but only appear that way from our point of view and are probably nowhere near each other in actual space. This is considered the most boring of all Messier objects and is located in the constellation Ursa Major. The stars are thought to be some 510 light years away.

Where it is:
Right ascension: 12h 22m 12.5s
Declination: +58° 4' 59"

Imaging the target: This is, without any doubt, the single most boring Messier object. It is two close stars, both almost the same brightness, and almost the same color, with no real field behind them. No field is probably a good thing since they would just blend in if there were. Why this was included in Messier's list is beyond me.

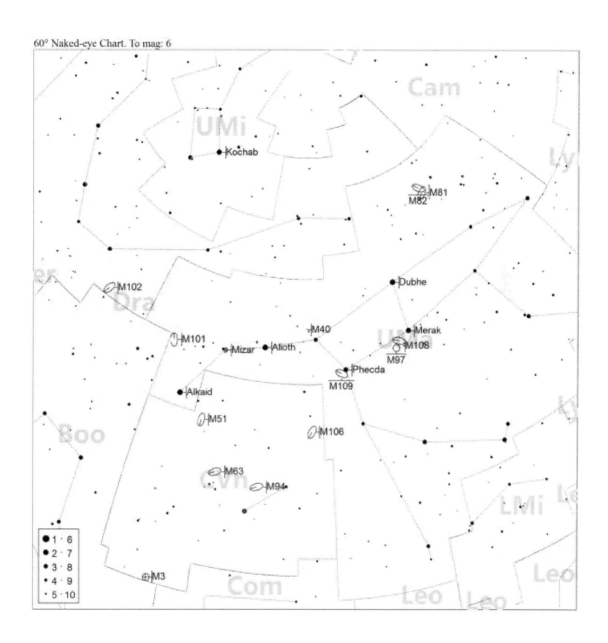

Messier 41:

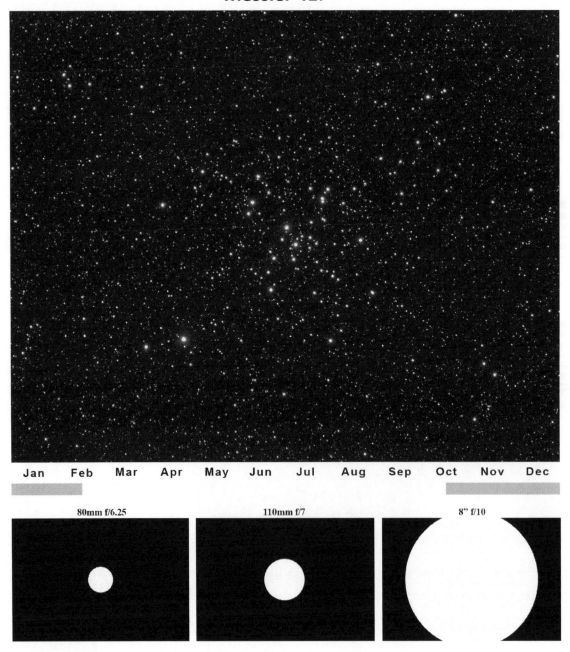

About the target: This open cluster is thought to be roughly 200 million years old and sits 2,300 light years away in the constellation Canis Major. The cluster spans some 25 light years across. It is possible that this cluster was identified by Aristotle in his book *Meteorologica* in 325BC although discovery is usually attributed to Giovanni Batista Hodierna sometime before the mid 1600s. M41 was added to Messier's list on the 16th of January, 1765.

Where it is:
Right ascension: 06h 46.0m
Declination: −20° 46′

Imaging the target: This is another semi boring open cluster in the Messier list. Its saving grace, like several others, is that if you shoot in color it has a few nice colored stars in the shot that help keep things a little interesting. Like several before it, this is a midrange exposure object, 240 seconds at ISO 800 for me, and responds well to only minor stretching. Leave your white point alone when processing to make sure that you keep as much star color as possible. Thermal noise is not that much of an issue here as this is a good target to shoot in the late fall months so temperatures are relatively cool.

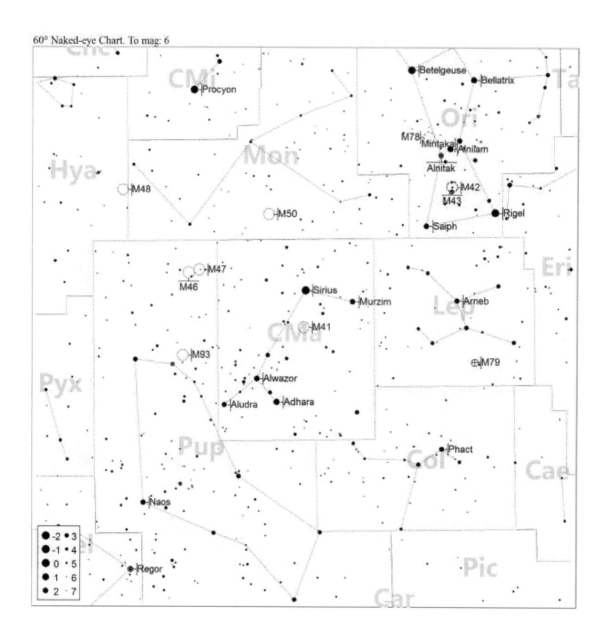

Messier 42: The Orion Nebula

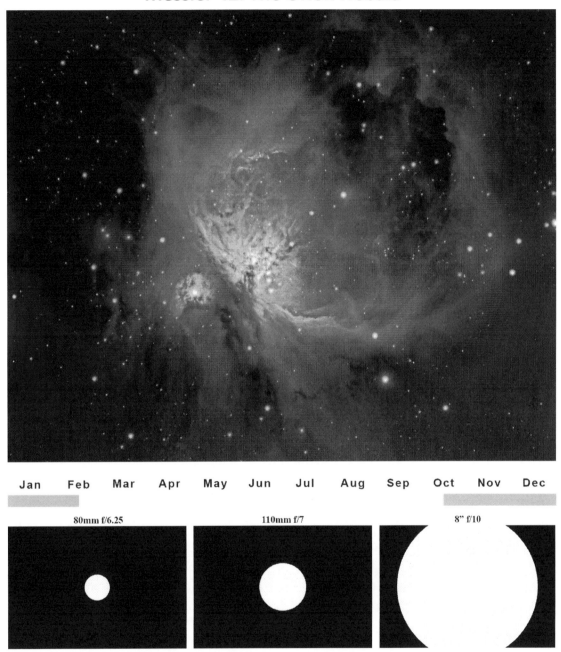

About the target: Very likely the most popular target in the Messier catalog and easily a naked eye object in even most suburban settings, this is the closest area of star formation to us at only 1,300 light years away. Discovered by French astronomer Nicolas-Claude Fabri de Peiresc on the 26TH of November, 1610 and added to this list on the 4TH of March, 1769 by Messier, it stretches some 25 light years across. In the very heart of the nebula lies the Trapezium Cluster: a cluster of stars easily visible with only a small telescope which shows six stars under relatively dark skies. The Trapezium was discovered on the 4TH of February, 1617 by Galileo Galilei.

Where it is:
Right ascension: 05h 35m 17.3s
Declination: −05° 23′ 28″

Imaging the target: This is most people's favorite target and probably the most photographed object in the Messier list. It is also probably the brightest being easily seen with the naked eye. Don't let that fool you. Good pictures are tough to get because parts of it are very bright, while other parts are very dim.

While all very short exposures, this image took 9 sets of images shot at 9 different exposures. The result of each set was then combined in an HDR program to generate the image that you see. Since it was cold when I shot the target, I was not really worried about thermal noise. The exposures I used at ISO 800 were 10 images at 5 seconds, 10 at 10 seconds, 10 at 15 seconds, 10 at 30 seconds, 10 at 45 seconds, 10 at 60 seconds, 10 at 90 seconds, 10 at 120 seconds and 10 at 180 seconds. I then took 20 darks at each exposure for stacking.

There was a lot of stretching in small amounts over several sessions using curves to try to bring out the really faint wisps of dust clouds. Take your time or you will overdo parts of it. Be very careful here; it is extremely easy to overstretch it. I like to do a small stretch, then copy the layer, and then do another small stretch, and so on. This makes sure that when I overdo it (not if, but when) I can easily go back without having to start over. I can also compare stretches by hiding and revealing layers.

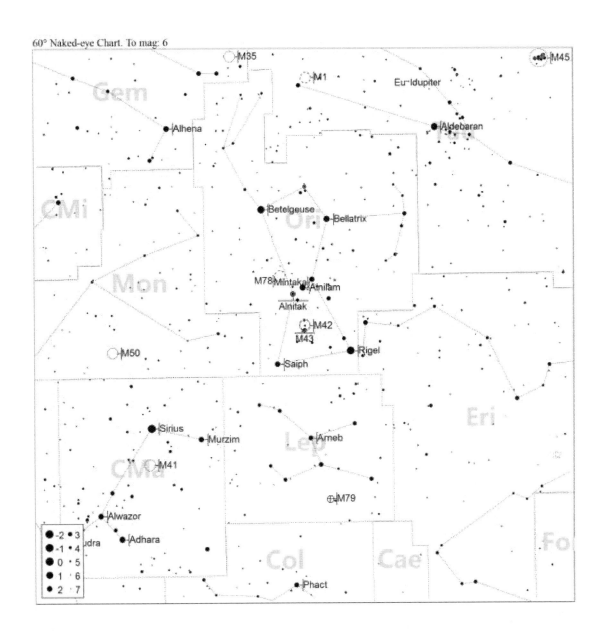

Messier 43: De Mairan's Nebula

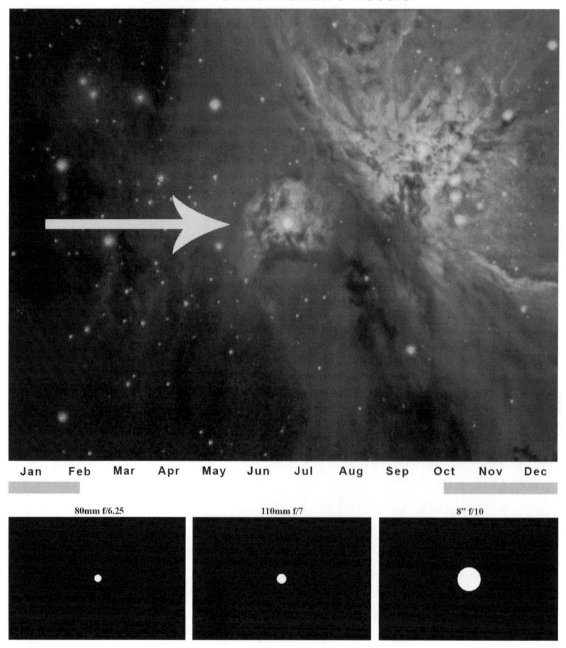

About the target: Even though this is a part of the Orion Nebula, M42, it is harder to see and separated from M42 by a dust lane. Discovered sometime around 1731 by French astronomer Jean-Jacques d'Ortous de Mairan, it sits some 1,300 light years away. Messier described M43 as "Position of the little star surrounded by nebulosity & which is below the nebula of the sword of Orion" on the 4TH of March, 1769.

Where it is:
Right ascension: 05h 35.6m
Declination: −05° 16′

Imaging the target: De Mairan's Nebula is really just a part of M42 and in medium to wide field of views should be treated as one single nebula. If you have something like an 8" SCT however you might want to try to add a 2x barlow and image just this part by itself. Regardless, the processing will remain the same as it too has a bright core, some medium brightness nebula and some very faint dust lanes.

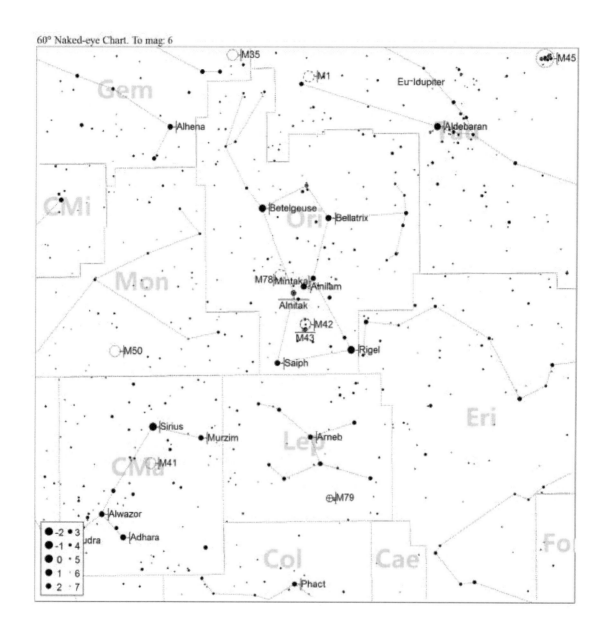

Messier 44: The Beehive Cluster

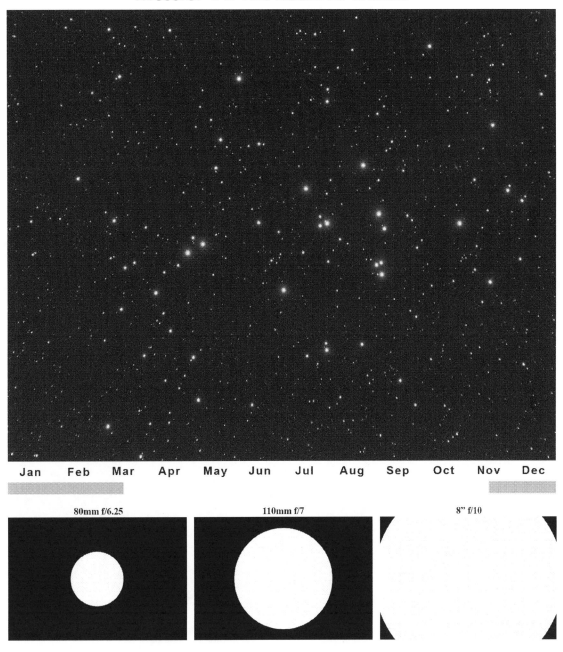

About the target: The constellation of Cancer is home to this open cluster which sits over 500 light years away. Known since ancient times, and possibly first cataloged by Hipparchus in 130 BC and again by Ptolemy around 130 AD, this 600 million year old cluster is large and bright and is easily seen with the naked eye from reasonably dark skies. In 1609 Galileo used his telescope to find 40 of the approximately 350 stars present in this cluster.

Where it is:
Right ascension: 08h 40.4m
Declination: 19° 41'

Imaging the target: The Beehive Cluster is one of the open clusters that is a lot more fun to look at than image. There are some pretty nice colors in the stars if you are shooting color but not that much of a background field. The members of the cluster stand out fairly well however. Like several before it, this is a midrange exposure object, 240 seconds at ISO 800 for me, and responds well to only minor stretching. Leave your white point alone when processing to make sure that you keep as much star color as possible. Thermal noise is not that much of an issue here as this is a good target to shoot in the winter months so temperatures are relatively cool.

60° Naked-eye Chart. To mag: 6

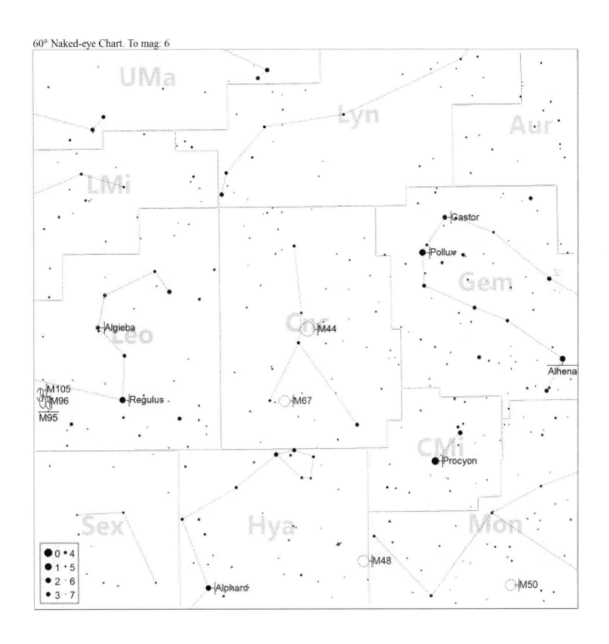

Messier 45: The Pleaides

About the target: The Pleiades, or Seven Sisters, has been known since recorded histories of objects in the skies have been kept. Early records from 1600 BC seem to mention the open cluster in Taurus written on the Nebra Sky Disk (an almost 8" disc of bronze depicting the sun, moon, stars and the Pleiades). The cluster has also been mentioned by Homer, in the Bible and many other sources down through history. Just over 400 light years distant this is without a doubt the brightest cluster of stars in the sky. The brightest of stars are surrounded by nebulosity which while not directly visible, is easily captured in images. Messier cataloged this on the 4TH of March, 1769.

Where it is:
Right ascension: 3h 47m 24s
Declination: +24° 7′

Imaging the target: This is what we wish all open clusters were like. Visually with no optical aids, with binoculars, with a telescope, or imaging, this is a beautiful object that can really make you smile. Around these quite bright stars (magnitudes up to 2.85) is a wonderful area of blue nebulosity. Unfortunately blue nebulosity is very dim compared to the stars that provide the light for it, so getting both the stars and the nebulosity can be a challenge.

I recommend shooting a short exposure, in my case 100 seconds at ISO 800, and shooting a lot of frames (30 at a minimum, preferably lots more). Be very careful here, it is extremely easy to overstretch it, take your time or you will make the stars blow out huge. I like to do small a stretch, then copy the layer, and then do another small stretch, and so on. This makes sure that when I overdo it (not if, but when) I can easily go back without having to start over. I can also compare stretches by hiding and revealing layers.

Have patience, this is a tough target to get the way you want.

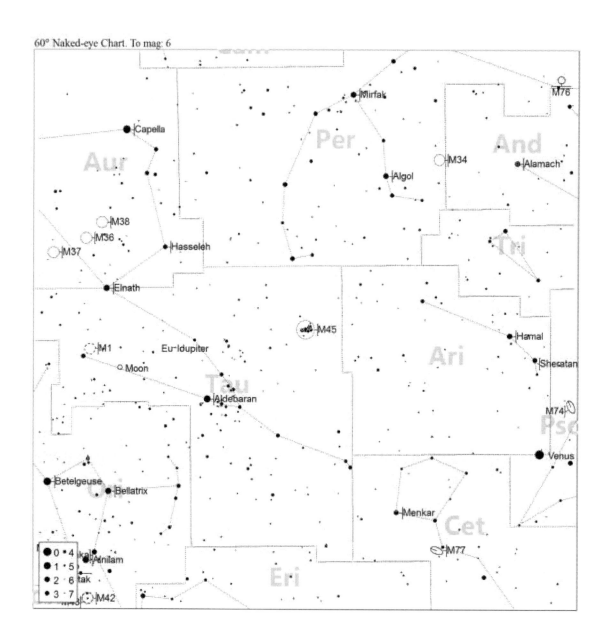

Messier 46:

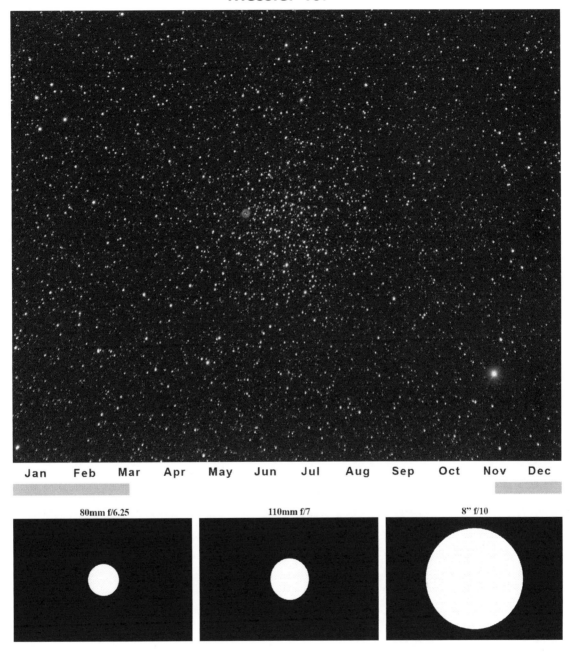

About the target: In the constellation Puppis, Messier discovered this open cluster on the 19TH of February, 1771. This 300 million year old cluster sits some 5,400 light years away with a diameter of approximately 30 light years. Looking close one can see NGC2438, a blue planetary nebula included, mixed in with the estimated population of 500 stars in this cluster. This most likely is an optical illusion with the planetary nebula somewhere between us and the cluster thus appearing to be a member.

Where it is:
Right ascension: 07h 41.8m
Declination: −14° 49′

Imaging the target: Now here is a open cluster that is deceptively interesting. Looking at it through a telescope will show nothing overly interesting, a little star color, especially from that bright big yellow guy on the lower right (HIP 37379), but not much else. Take a picture at medium exposure (150-240 seconds for me) and you get a little blue halo around one star just outside the center to the left. Try as you might, you can't get rid of that halo. I even moved the scope a little to see if it was on my telescope or camera sensor!

Nope, that little guy is a planetary nebula (PLN231+4.2) and is a nice little surprise for your image. It is so bright you really don't even have to do any stretching which is probably a good thing considering how bright that star in the lower right is.

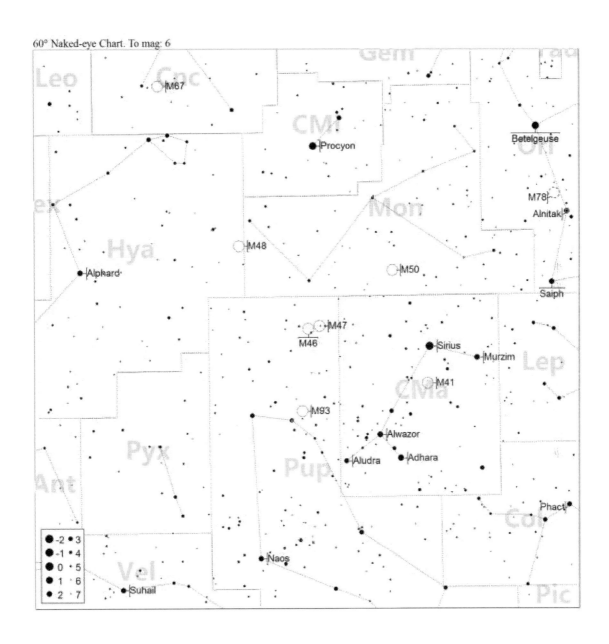

Messier 47:

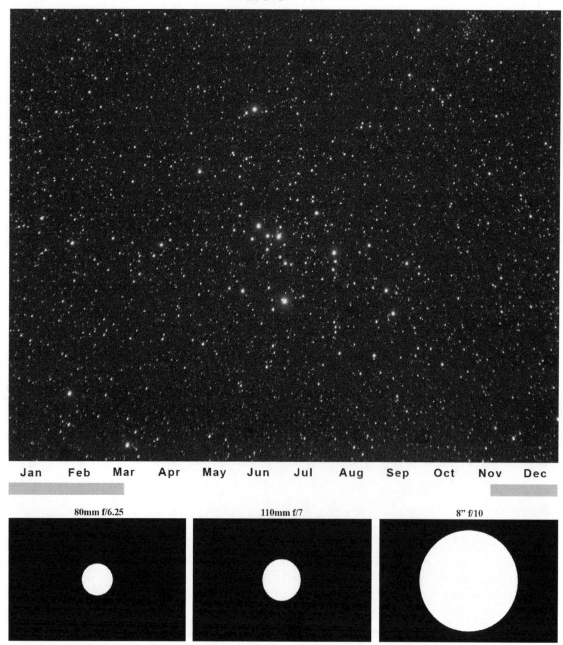

About the target: This open cluster is right next door to M46, and if you think you see it all the way to the left of this image you are not alone, but unfortunately incorrect. The cluster to the far left is NGC2423, another open cluster. M47 sits in the constellation of Puppis and is roughly 1,600 light years distant and contains 50 or more stars. It was discovered by Charles Messier on February 19TH, 1771.

Where it is:
Right ascension: 07h 36.6m
Declination: -14° 30'

Imaging the target: Again, this is another semi boring open cluster in the Messier list. Its saving grace, like several others, is that if you shoot in color it has a few nice colored stars in the shot that help keep things a little interesting. Like several before it, this is a midrange exposure object, 240 seconds at ISO 800 for me, and responds well to only minor stretching. Leave your white point alone when processing to make sure that you keep as much star color as possible. Thermal noise is not that much of an issue here as this is a good target to shoot in the late fall months so temperatures are relatively cool.

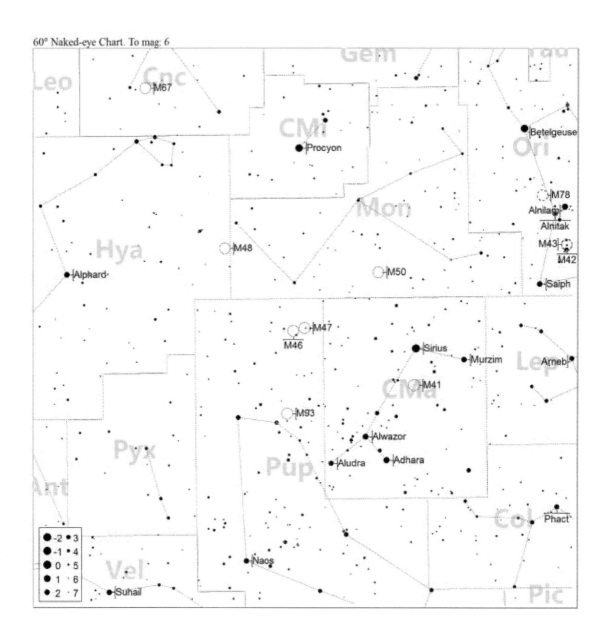

Messier 48:

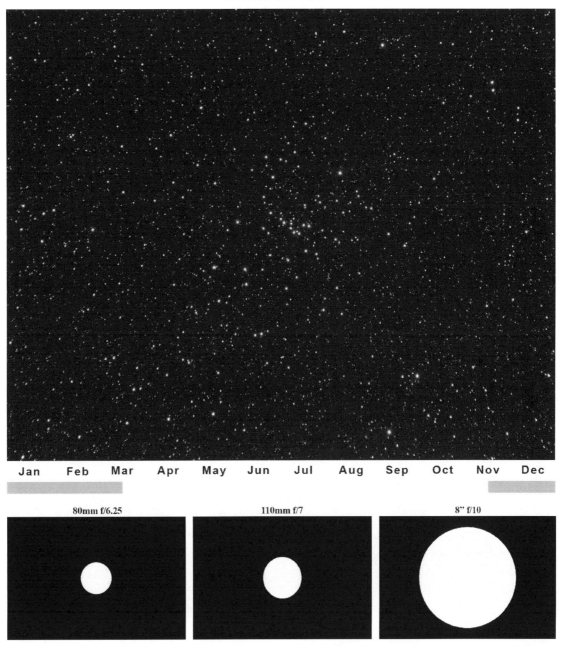

| Jan | Feb | Mar | Apr | May | Jun | Jul | Aug | Sep | Oct | Nov | Dec |

80mm f/6.25 110mm f/7 8" f/10

About the target: On the 19th of February, 1771, Messier discovered this cluster in the constellation of Hydra, writing in his notes "Cluster of very small [faint] stars, without nebulosity". Some 300 million years old and 1,500 light years away this cluster is reasonably easy to spot with the naked eye from a semi-dark site. The cluster contains some 80 or more stars that span some 23 light years in diameter.

Where it is:
Right ascension: 08h 13.7m
Declination: -05° 45'

Imaging the target: And again, this is another semi boring open cluster in the Messier list. Its saving grace, like several others, is that if you shoot in color it has a few nice colored stars in the shot that help keep things a little interesting. Like several before it, this is a midrange exposure object, 240 seconds at ISO 800 for me, and responds well to only minor stretching. Leave your white point alone when processing to make sure that you keep as much star color as possible. Thermal noise is not that much of an issue here as this is a good target to shoot in the late fall months so temperatures are relatively cool.

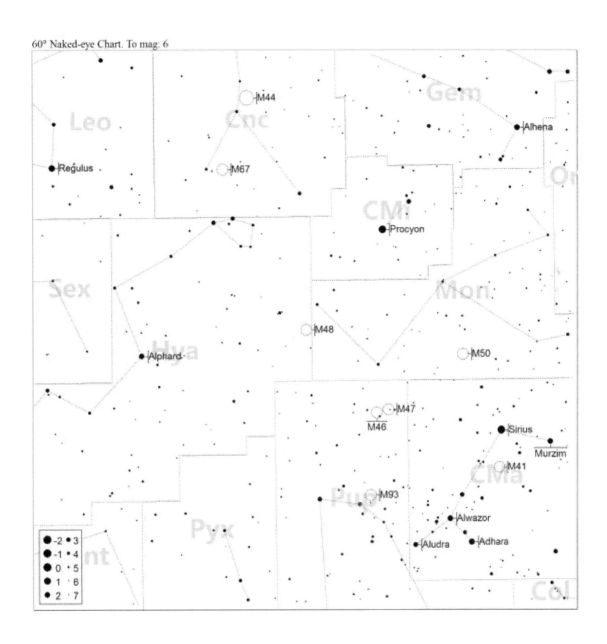

Messier 49:

Jan Feb Mar Apr May Jun Jul Aug Sep Oct Nov Dec

80mm f/6.25　　　　110mm f/7　　　　8" f/10

About the target: This elliptical galaxy is roughly 49 million light years away and appears in the constellation of Virgo and x-rays emitted by the central core suggest it contains a supermassive black hole. If you look to the left of M49 and just a little down you may see that smudge, which is another galaxy NGC4465. You may not however expect that the two stars between M49 and NGC4465 are not both stars! One is indeed the star GSC874:941 (the one closest to NGC4465) but the other is actually yet another galaxy, NGC4467.

Where it is:
Right ascension: 12h 29m 46.7s
Declination: +08° 00′ 02″

Imaging the target: Now we come to an elliptical galaxy and it is not a lot of fun to shoot because you will never get any detail out of it. Instead, what you may want to concentrate on is going very deep with long exposures and seeing how many other galaxies you can find in the same shot. For example, in the image I presented here of this target, there are several more galaxies in the image other than M49. Unfortunately we will never get any detail out of them either, but it is fun to see how many you can get.

I suggest long exposures of 400 or more seconds for my setup at ISO 800.

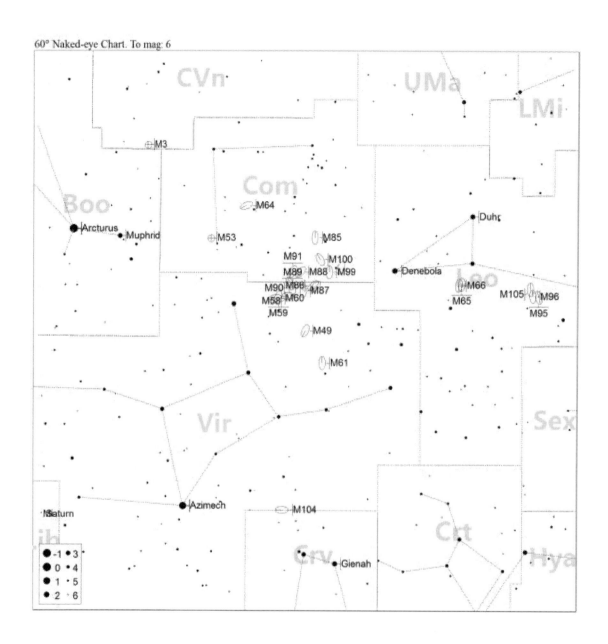

Messier 50:

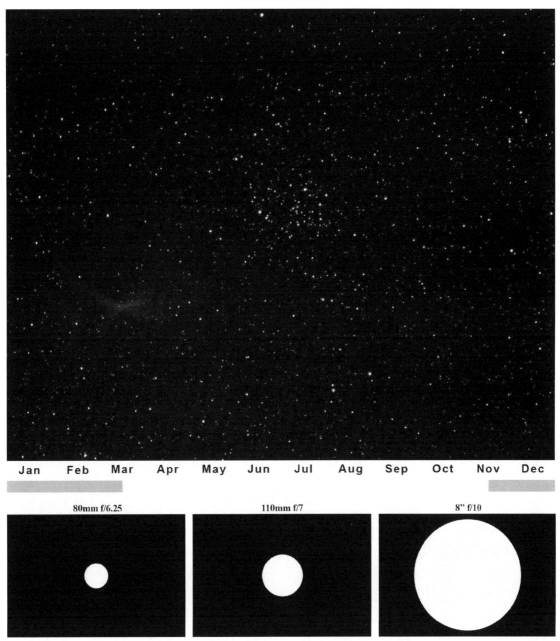

About the target: Thought to be discovered around 1711 by Giovanni Domenico Cassini and later separately discovered by Messier on the 5th of April, 1772, this cluster in the constellation of Monoceros appears to be heart shaped. Spanning some 20 light years across this cluster is approximately 3,200 light years away. The latest estimates put approximately 200 stars in the central region of the cluster.

Where it is:
Right ascension: 07h 03.2m
Declination: −08° 20′

Imaging the target: Much like M46 this is a deceptively interesting cluster. Visually, not too interesting. Once you put several medium length (150-240 seconds at ISO 800) exposures together and start stretching however you may start to see a smudge over on the left just below center, this is the planetary nebula NGC2316. In addition, you can get some fairly nice star colors out of this area if you are shooting color.

I recommend shooting around 240 second exposures for my equipment and quite a few of them, say 25 or more. Since this is up in the winter, thermal noise should not be an issue. Leave your white point alone and do an initial levels stretch to catch a glimpse of the nebula without blowing out the star colors, then use curves to bring it out even more. You should be able to bring out much more than you may see in my image, I really need to go back to this target and do a better job.

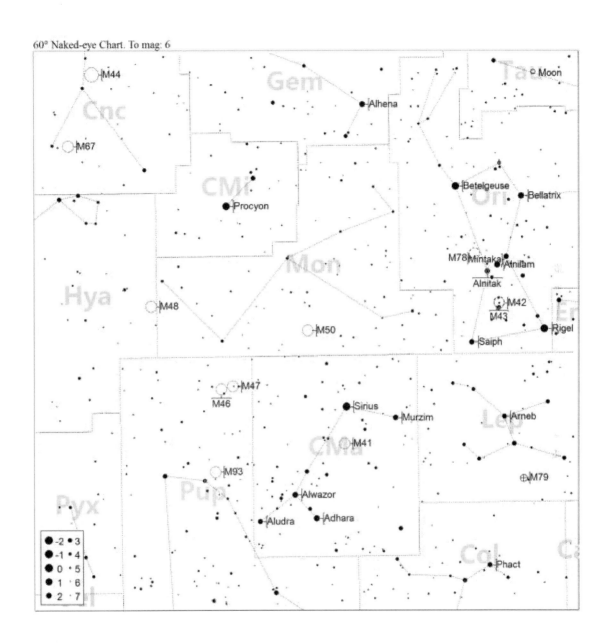

Messier 51: Whirlpool Galaxy

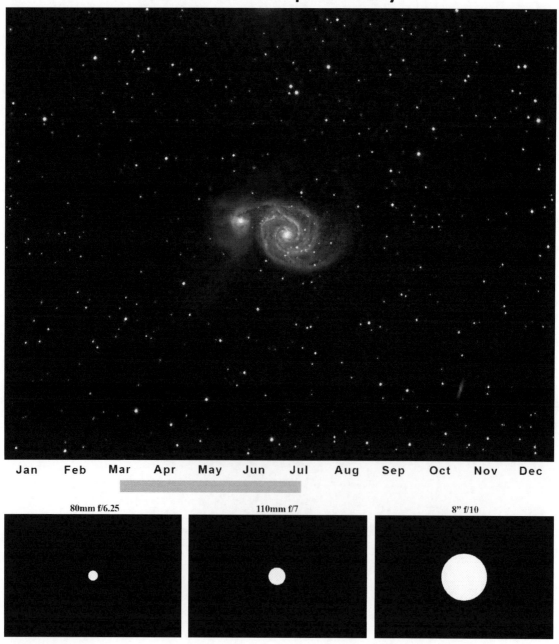

Jan   Feb   Mar   Apr   May   Jun   Jul   Aug   Sep   Oct   Nov   Dec

80mm f/6.25   110mm f/7   8" f/10

About the target: Messier discovered the Whirlpool Galaxy or Rosse's Galaxy, the larger of the two galaxies that are collectively considered M51, in the constellation of Canes Venatici on the 13th of October, 1773. Its "Rosse's Galaxy" name stems from Irish astronomer William Parsons, the 3rd Earl of Rosse, who in 1845 painted a very detailed likeness of the galaxies. The smaller galaxy in this pair was discovered on the 21st of March, 1781 by French astronomer Pierre Méchain. Several million years ago the large spiral galaxy "ran into" the smaller dwarf galaxy (NGC5195), ripping off some of its material to add to its own. Sitting some 25 million light years away, this amazing object is one you have to see. So far there have been two recorded supernovae in M51; SN 2005cs and SN 2011dh.

Where it is:
Right ascension: 13h 29m 52.7s
Declination: +47° 11' 43"

Imaging the target: One of the most interesting objects in the Messier catalog is actually two objects in one, two galaxies that is. The closer you can get, the more interesting this object will be, right up to filling the image. The down side of doing that of course is your polar alignment and guiding need to be dead on when using a long focal length scope like a 8" SCT or more.

As for shooting this target you need to go very deep, in my case lots of 480 second ISO 800 images stacked. It is still pretty cool when this target is up so thermal noise is not too much of an issue but the longer exposures still introduce a lot of extra noise which is why I suggest lots of images. Once you have that done, stretching this is challenging as you need to stretch the outer arms without blowing the core. An initial stretch with levels should be sufficient to get you started and then switch to curves and stretch in small increments to maximize the amount of detail you can pull out.

Galaxies like this one may require something you haven't heard me talk about yet and that is sharpening. When applying sharpening to this you need to do it carefully and with a mask so you are only sharpening the galaxy and not the black of space around it. The reasoning here is that if you try to sharpen everything you can really increase the noise levels in the background which will make the overall image look very grainy.

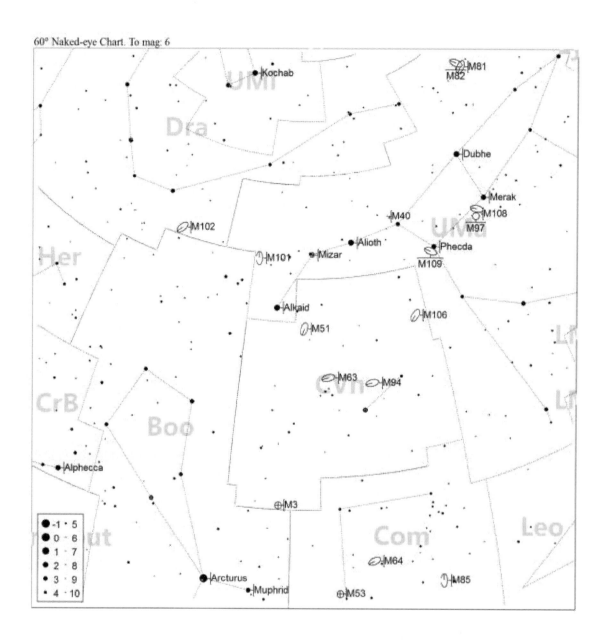

Messier 52:

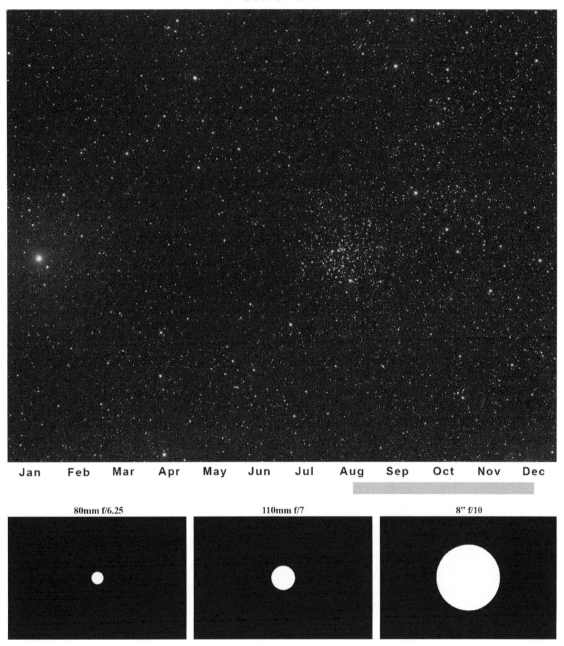

About the target: This open cluster appears in the constellation Cassiopeia some 5,000 or so light years away and contains roughly 200 stars. Cataloged by Messier on the 7th of September, 1774, this cluster requires a pair of binoculars or small telescope to see. The almost 200 members of this cluster spans some 19 light years across. That bright orange star to the left is HIP 115590, a magnitude 4.96 star.

Where it is:
Right ascension: 23h 24.2m
Declination: +61° 35′

Imaging the target: This open cluster is a little less boring than most, primarily because of the entire field of view and framing. Note that the central portion of the cluster is off to the right just a little, and then to its left is an area with far fewer stars than on its right, this "negative space" is then interrupted by the bright yellow star (HIP 115590). To me, this juxtaposition makes the image.

To shoot this target I used a medium exposure of 240 seconds at ISO 800 and was very sparing on the stretching as to not blow out the bright star on the left. I did have to deal with a substantial amount of thermal noise as this is shot in the middle of a Texas summer. Looking back, I wish I had shot many more frames to help with the noise, something like 30 or so would have improved things.

60° Naked-eye Chart. To mag: 6

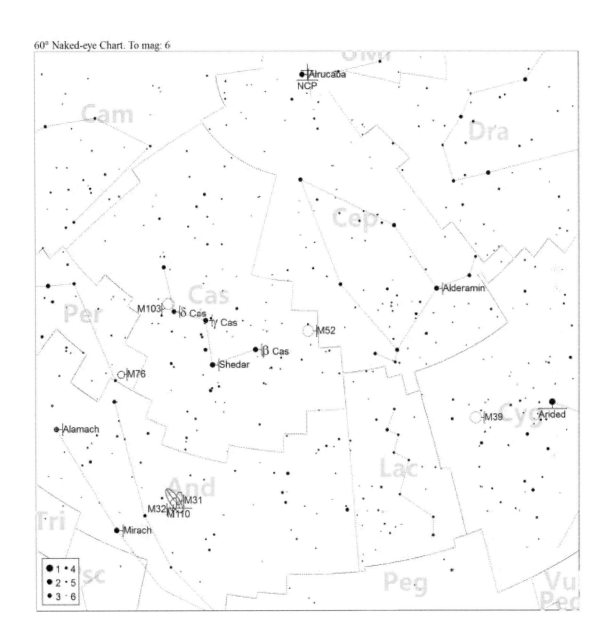

Messier 53:

Jan Feb Mar Apr May Jun Jul Aug Sep Oct Nov Dec

80mm f/6.25   110mm f/7   8" f/10

About the target: In the constellation Coma Berenices is this globular cluster discovered by German astronomer Johann Elert Bode on the 3rd of February, 1775. It is approximately 58,000 light years away and is thought to be some 12.6 billion years old. Messier cataloged it separately on the 26th of February, 1777.

Where it is:
Right ascension: 13h 12m 55.25s
Declination: +18° 10' 05.4"

Imaging the target: M53 is another reasonably tricky globular as the core is very dense and bright, while the spread is relatively dim. My image here does a poor job at showing off the spread and is a little too bright in the core (hindsight and all that). The good news is that this globular has a rather impressive spread if you can get it to come out.

I would recommend that you take one set of images at around 150 seconds at ISO 800 (for my equipment) for the core and be very sparing on the stretching. Then take a second set of images for the spread at somewhere around 240-300 seconds at ISO 800 and you can be a little more aggressive on the stretching midpoints but be careful to only blow out the core and not the other stars around the globular. You can then combine these images using HDR software or layers to create one interesting image.

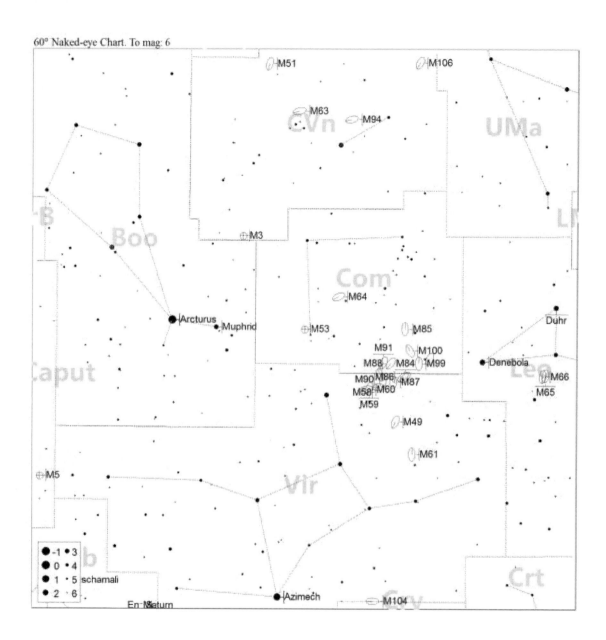

Messier 54:

About the target: Discovered by Charles Messier on the 24th of June 1778 this globular cluster in Sagittarius was originally thought to be part of our galaxy, however in 1994 it was determined that the Sagittarius Dwarf Elliptical Galaxy is more likely its true home. At a distance of 87,000 light years spanning over 300 light years across, this is truly a massive globular cluster even though it appears fairly small to us. This cluster can easily be seen in small amateur telescopes but takes a professional telescope to resolve into stars because of its extreme density.

Where it is:
Right ascension: 18h 55m 03.33s
Declination: −30° 28′ 47.5″

Imaging the target: This small dense globular is going to be very difficult to do much with in a scope with a wider field of view but may be pretty interesting in a scope with a longer focal length such as an 8" SCT. The core is fairly bright but can somewhat be managed through medium length exposures around 240 seconds at ISO 800 on my equipment and then being very subtle with stretching. A nice feature is that the background field of stars is pretty nice so even in shorter focal length scopes you have something to work with. If you shoot in color try very hard to preserve the star colors to give the image that extra little pop.

60° Naked-eye Chart. To mag: 6

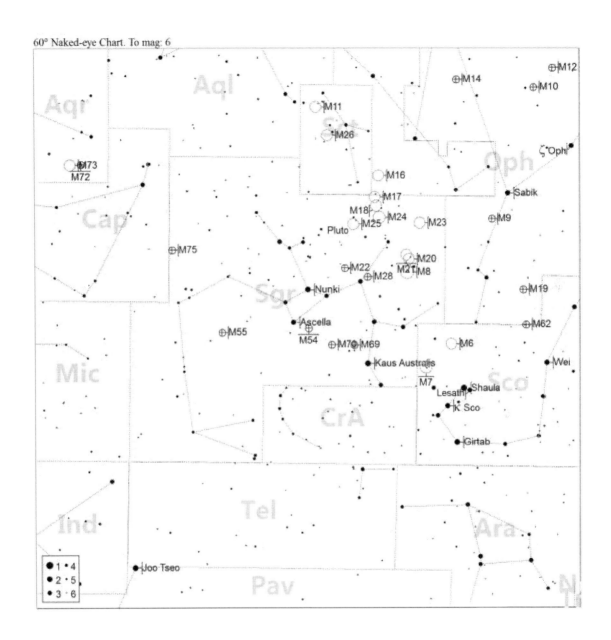

Messier 55:

Jan Feb Mar Apr May Jun Jul Aug Sep Oct Nov Dec

80mm f/6.25   110mm f/7   8" f/10

About the target: This globular cluster is in the constellation Sagittarius and was discovered by the French astronomer Nicolas Louis de Lacaille on the 16th of June, 1752. It contains over 50 variable stars and is some 17,500 light years away. Even though this is a large globular cluster, spanning some 100 light years, it has an unusually low number of known variable stars, possibly as few as 5. Messier cataloged it on the 24th of July, 1778.

Where it is:
Right ascension: 19h 39m 59.71s
Declination: −30° 57′ 53.1″

Imaging the target: This is a really fun globular to play with. It is dense enough to be quite obviously a globular cluster, yet loose enough to make the core very detailed. The penalty for that is that there is not much of a spread. Fortunately this is a relatively easy cluster to image but you need to go a little deeper than you might expect. I used 300 seconds at ISO 800 on my equipment and it seems to have worked well. The core resolves clearly to lots of tiny stars, partly due to their color difference. The background field is nicely populated with a multitude of colors which offsets the cluster well. Watch your stretching to maintain those star colors.

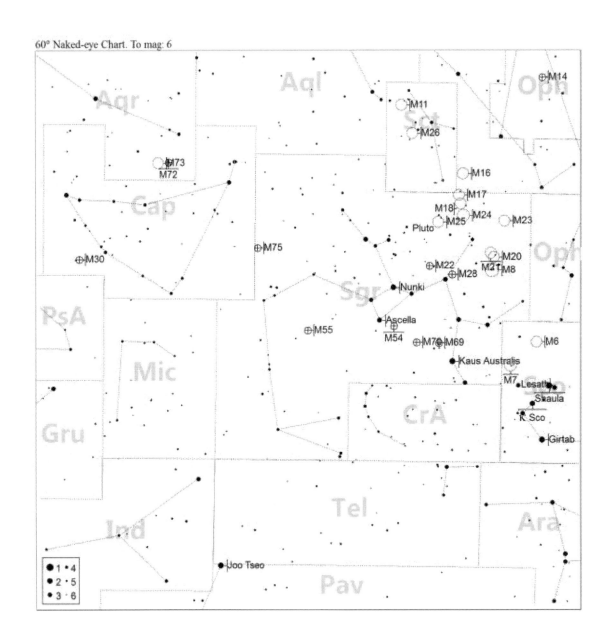

Messier 56:

About the target: On the 19th of January, 1779, Charles Messier discovered this globular cluster in Lyra calling it "Nebula without stars, having little light." Approximately 33,000 light years distant and 82 light years in diameter this small, loosely packed globular can be challenging to find in smaller telescopes. Messier was not stretching the truth when he said "having little light," even the brightest stars in the cluster are only 13th magnitude.

Where it is:
Right ascension: 19h 16m 35.57s
Declination: +30° 11' 00.5"

Imaging the target: This almost is not a globular cluster at all, but rather a dense open cluster. Where do they draw that line anyway? To me, it looks somewhat like a spider.
Capturing this is fairly straight forward; medium length exposures of 150 seconds at ISO 800 worked well for me along with minimal stretching to keep the detail in the cluster as well as the color in the stars in the field. This is especially important for the two stars in the lower left side which are yellow and blue if you shoot color and really offset each other well. This is probably the whole area's best feature unfortunately.

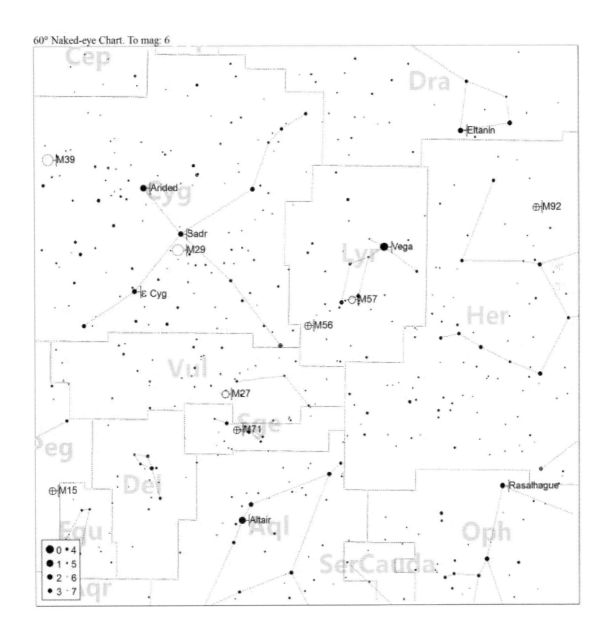

Messier 57: The Ring Nebula

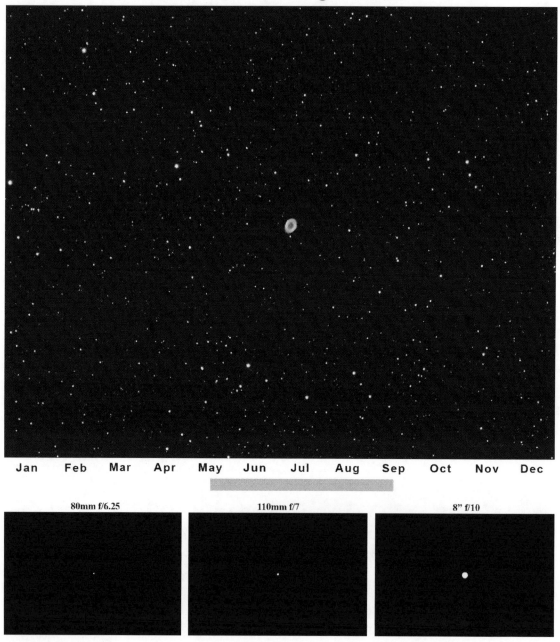

About the target: This planetary nebula in Lyra is a ring of ionized gas having been shot out from the surface of a red giant star while transforming into a white dwarf. Discovered in 1779 by French astronomer Antoine Darquier de Pellepoix it is just over 2,000 light years from us. Being very small at only 3 light years across, it is virtually impossible to see with handheld binoculars but in larger scopes is one of the few targets I have seen in full color, a beautiful blue.

Where it is:
Right ascension: 18h 53m 35.079s
Declination: +33° 01' 45.03"

Imaging the target: Here is a fun object no matter what focal length telescope you have, the Ring Nebula. Shoot fairly deep for this target at something like 300 seconds at ISO 800 for my setup and play with your stretching. If you look very close you may be able to see the white dot of a central star in the center of the nebula which could be the star that blew off its shell creating this nebula. Be careful in your stretching that you do not take the second most outer ring to white as I tend to do. It can be very difficult in a smaller scope to isolate the different parts of the shell for stretching. This gets a little easier with a longer focal length scope. If shooting color, be sure to preserve the outermost red ring as it adds a lot of kick to the image.

Messier 58:

About the target: On the 15th of April, 1779, Charles Messier discovered this barred spiral galaxy (essentially a spiral galaxy with a bar shaped central region) 62 million light years from us in the constellation of Virgo. This is one of the many galaxies in the Virgo "super cluster". This cluster of galaxies is so packed that Messier discovered three galaxies including this one in a single day, the other two being M59 and M60 of course.

Where it is:
Right ascension: 12h 37m 43.5s
Declination: +11° 49' 05"

Imaging the target: This barred spiral galaxy can actually give you some detail even in shorter focal length scopes such as mine. To do it right, and get more detail than my image shows, you will need to shoot deep and a lot of lights. For me, I would say 480 second at ISO 800 exposures at a minimum of 30-40 images stacked. An initial levels stretch should be enough to get some of the arms to show up and then use curves from there.

Those of you with good eyes may have seen a lot of elongated stars in my image and think my tracking was really off. While that is always possible (and here I think it was a little off), if you look closer you will note that there are several perfectly round stars as well. So how is it possible to have perfectly round stars in the same frame as elongated stars? I will give you a hint, there is probably somewhere near 50 other galaxies in this shot besides M58, cool huh?

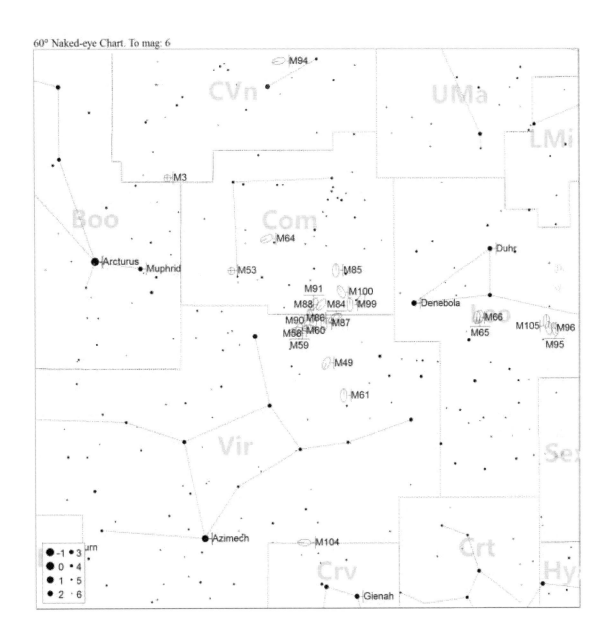

Messier 59:

About the target: Discovered by German astronomer Johann Gottfried Koehler on the 11th of April, 1779, this elliptical galaxy sits some 60 million years from us appearing in the constellation Virgo. Four days later, Messier discovered the galaxy and listed it in his catalog. This faint galaxy is part of the Virgo SuperCluster. It is thought that there are some 1,900 globular clusters in this galaxy.

Where it is:
Right ascension: 12h 42m 02.3s
Declination: +11° 38′ 49″

Imaging the target: Here is another fairly boring elliptical galaxy. The saving feature of this image is that looking closely, you can see several other galaxies as well. Shooting deep, for me somewhere near 480 seconds at ISO 800 and careful stretching can make quite a few galaxies jump out of this image. In my opinion, the best way to shoot these types of targets is when you can get as many targets in one frame as possible such as the picture I included near the front of this book in the section on the Virgo SuperCluster.

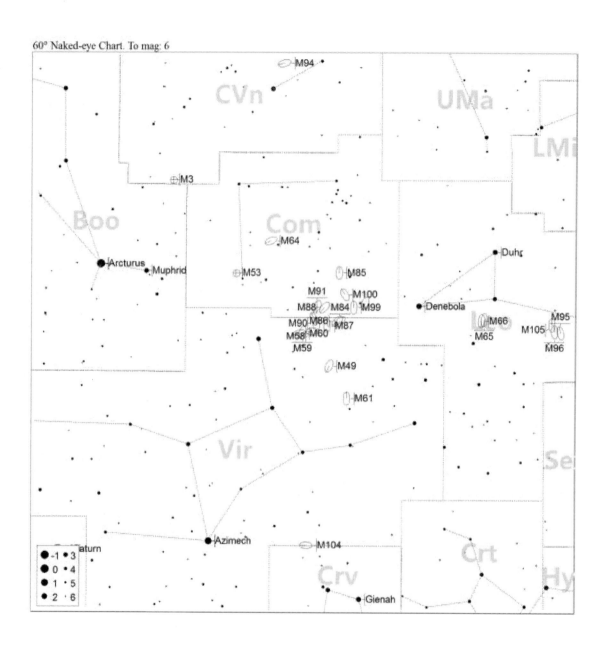

Messier 60:

About the target: Discovered by German astronomer Johann Gottfried Koehler on the 11th of April, 1779, this elliptical galaxy sits some 55 million years from us appearing in the constellation Virgo. Four days later, Messier discovered the galaxy and listed it in his catalog. This faint galaxy is part of the Virgo SuperCluster. Calculations say that this galaxy is some 120,000 light years across although to us, it appears very small.

Where it is:
Right ascension: 12h 43m 39.6s
Declination: +11° 33' 09"

Imaging the target: This is pretty much a repeat of M59, so I will repeat. Here is another fairly boring elliptical galaxy. The saving feature of this image is that looking closely, you can see several other galaxies as well. Shooting deep, for me somewhere near 480 seconds at ISO 800 and careful stretching can make quite a few galaxies jump out of this image. In my opinion, the best way to shoot these types of targets is when you can get as many targets in one frame as possible such as the picture I included near the front of this book in the section on the Virgo SuperCluster.

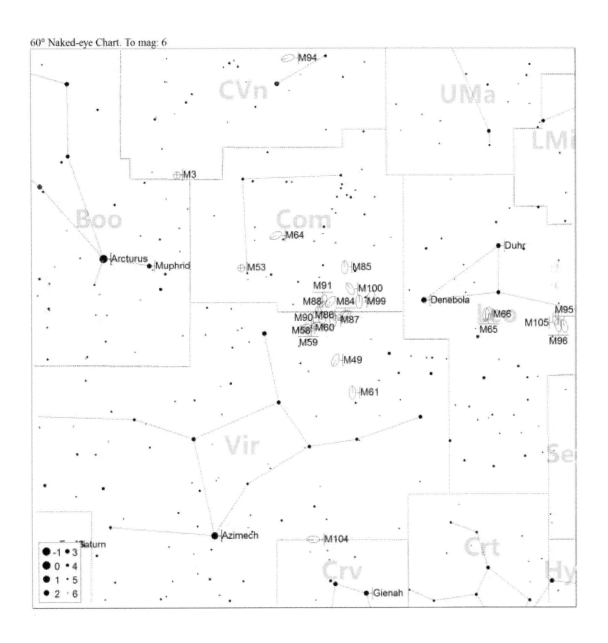

Messier 61:

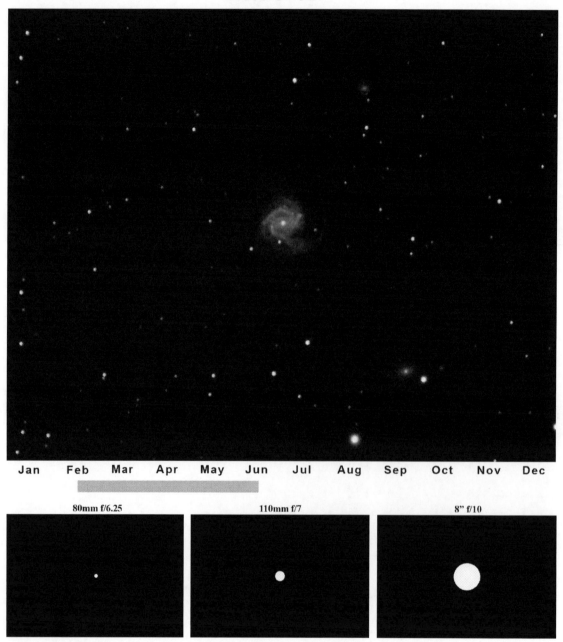

Jan Feb Mar Apr May Jun Jul Aug Sep Oct Nov Dec

80mm f/6.25   110mm f/7   8" f/10

About the target: This spiral galaxy is part of the Virgo Cluster of galaxies and was discovered by Italian astronomer Barnabus Oriani on May 5th, 1779, six days before Messier cataloged it. This galaxy is unique in that there have been six supernovae recorded in the past 100 years including two in the 21st century (SN 2008in, SN 2006ov). Calculations show this galaxy to span some 100,000 light years in diameter.

Where it is:
Right ascension: 12h 21m 54.9s
Declination: +04° 28′ 25″

Imaging the target: This is a frustrating little guy as I have nowhere near the amount of focal length I need to get a reasonable image of this target. I would guess you need at least 2000mm, something akin to an 8″ SCT or even more. Once you have that, take nice long exposures, as long as you can get away with is where I would start looking. Then I would take 30-40 of them for starters, and very carefully use curves to stretch it out. Lastly, I would create a mask and sharpen just the galaxy. Fortunately if you can do all that, thermal noise should not be too much of an issue as this is a late winter target.

60° Naked-eye Chart. To mag: 6

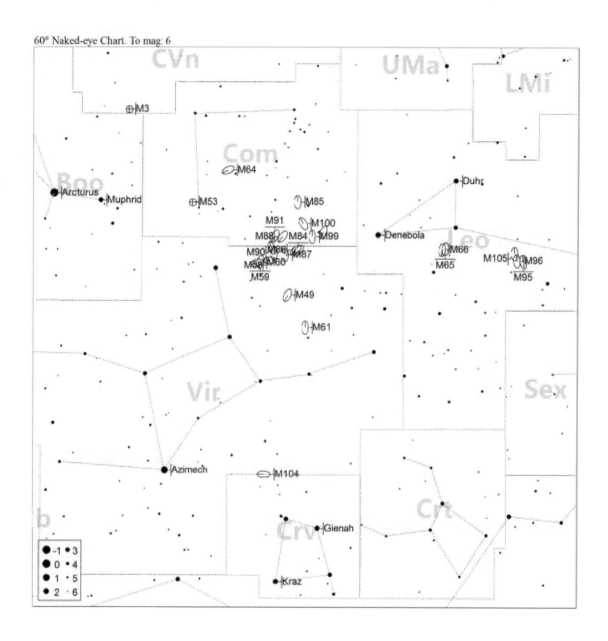

192

Messier 62:

About the target: This globular cluster in Ophiuchus was discovered by Charles Messier on the 7th of June, 1771, even though his logs show a date of the 4th of June, 1779. This was because of a problem with the accuracy of the position of the observation. At 22,000 light years distant with a diameter of almost 100 light years, it appears relatively small and very dense. Looking closely one can also see it is not very round and indeed seems to have been deformed by the tidal forces present since it is relatively close to the Galactic center.

Where it is:
Right ascension: 17h 01m 12.60s
Declination: −30° 06′ 44.5″

Imaging the target: This dense little globular has a larger than I would have expected spread and a wonderful background field of stars. The central core however is very dense and I managed to blow it out very quickly. The only shot I have ever seen where the core was not blown out was taken by the Hubble space telescope so I suppose I will have to live with it blown out.

Getting the spread and background field is pretty straight forward, a medium exposure set such as 240 second ISO 800 exposures worked quite well. It is starting to warm up a little when this target is high in the sky so you might want to take quite a few shots to help keep thermal noise minimal.

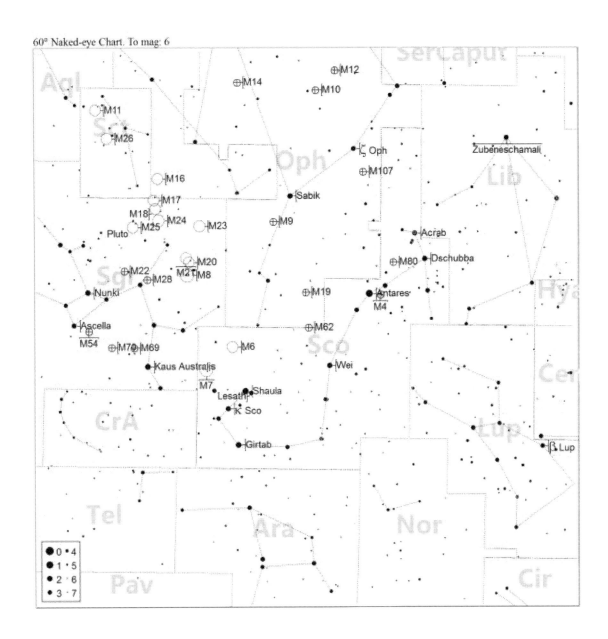

Messier 63: The Sunflower Galaxy

About the target: Close to M51 (the Whirlpool Galaxy) sits this spiral galaxy in the constellation Canes Venatici roughly 37 million light years away. It was first discovery by French astronomer Pierre Méchain on the 14th of June, 1779 and included in Messier's catalog that same day. Later around 1850, Irish astronomer William Parsons, the 3rd Earl of Rosse, noted the spiral appearance which classified it as a spiral galaxy.

Where it is:
Right ascension: 13h 15m 49.3s
Declination: +42° 01′ 45″

Imaging the target: The Sunflower Galaxy is a very interesting target that has a lot of detail if you take the time and effort to coax it out. The target is fairly bright so medium length exposures should do fine, 240 seconds at ISO 800 for me. Small targets with lots of details means lots of shots so I would start somewhere around 30 and see where that got me.

Stretching will have to be carefully done as the target is relatively small and has multiple sets of details you will want to bring out. If shooting in color, the outer blue ring will be the hardest to reveal without blowing out the central white areas. This of course is best accomplished with curves instead of levels and you might even want to go so far as to isolate the blue channel and use a mask to pull that out.

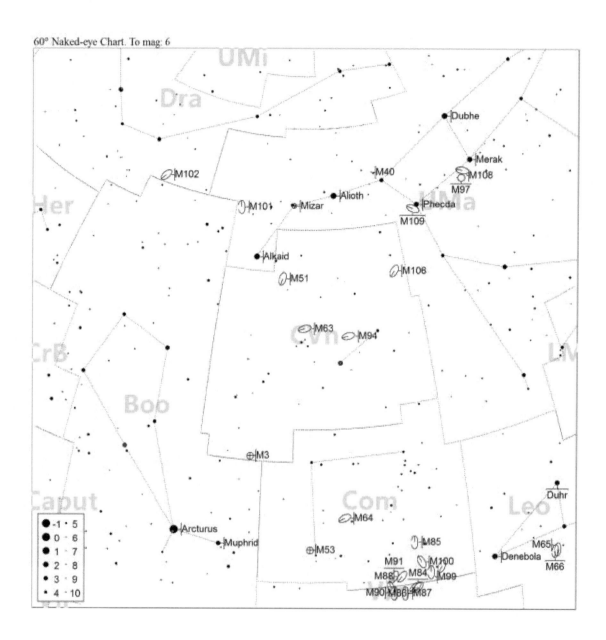

Messier 64: The Black Eye Galaxy

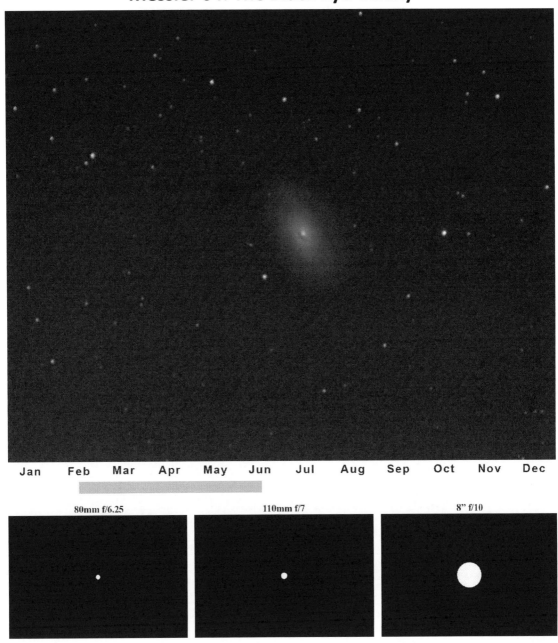

About the target: In the constellation Coma Berenices is this spiral galaxy discovered by English astronomer Edward Pigott in March of 1779 sitting some 24 million light years away. Note the black dust lane running under the left side of the central core making it appear to have a "black eye", hence the name. Messier cataloged this galaxy on the 1st of March, 1780. Unlike most galaxies, this one has counter rotating inner and outer areas which is most likely due to a collision, and absorption, of another neighboring galaxy.

Where it is:
Right ascension: 12h 56m 43.7s
Declination: +21° 40' 58"

Imaging the target: I hate to do this but M63 and M64 have nearly identical shooting and processing requirements. The Black Eye Galaxy is a very interesting target that has a lot of detail if you take the time and effort to coax it out. The target is fairly bright so medium length exposures should do fine, 240 seconds at ISO 800 for me. Small targets with lots of details means lots of shots so I would start somewhere around 30 and see where that got me. Stretching will have to be carefully done as the target is relatively small and has multiple sets of details you will want to bring out. If shooting in color the outer blue ring will be the hardest to reveal without blowing out the central white areas. This of course is best accomplished with curves instead of levels and you might even want to go so far as to isolate the blue channel and use a mask to pull that out.

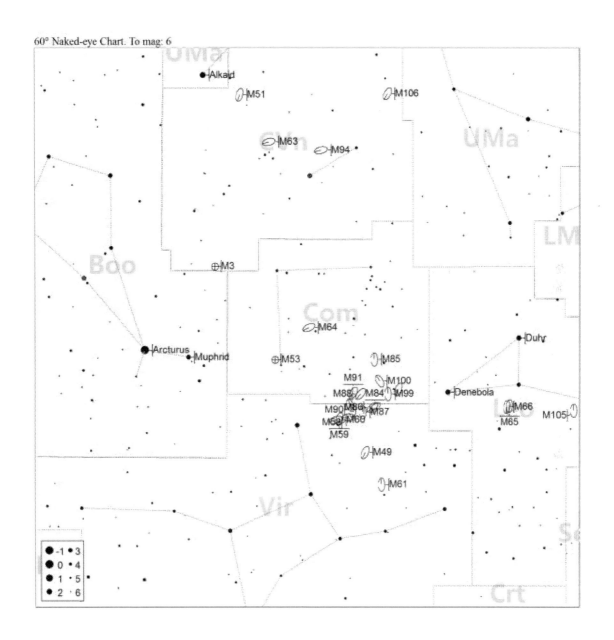
60° Naked-eye Chart. To mag: 6

Messier 65: The Leo Triplet

About the target: M65 is at the upper right, M66 at the lower right (both are intermediate spiral galaxies), and NGC 3628 (an unbarred spiral galaxy) is to the left. These three are what is commonly called the Leo Triplet because they sit in the constellation Leo. M65 is an intermediate spiral galaxy approximately 35,000,000 light years away as is NGC3628, however M66 is approximately 36 million light years away.

Where it is:
Right ascension: 11h 18m 55.9s
Declination: +13° 05′ 32″

Imaging the target: Because of my field of view, I always shoot M65 and M66 in the same frame. In fact, I always include NGC 3628 in the field as well. This takes an interesting shot of a small galaxy and turns it into something really special in my opinion. These three are commonly called the Leo Triplet.

These three little galaxies are not only beautiful and show lots of detail, they are also remarkable in that the same exposure can be used on all three. This makes them perfect to sit in the same frame. To extract that much detail from such little objects I used 32 images at 200 seconds at ISO 800 and 25 darks. I would really like to go back and shoot deeper to get a better background field and then merge this image with that one as the background here is not very spectacular. This is high in the sky in winter so thermal noise was not an issue, even with these little targets.

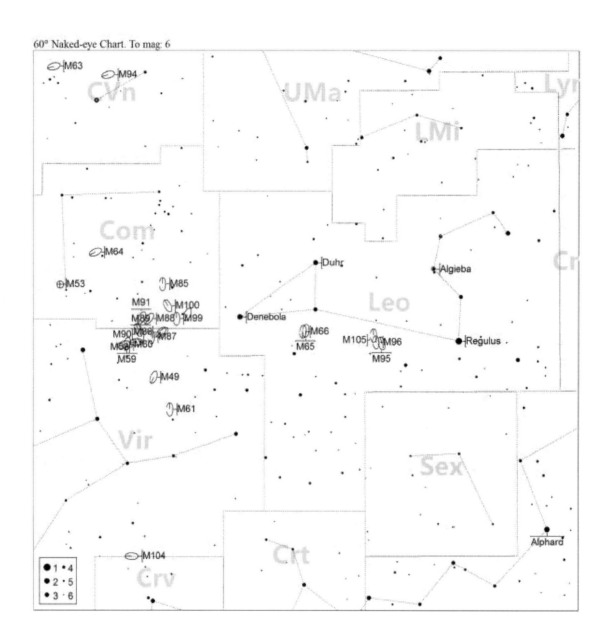

Messier 66: The Leo Triplet

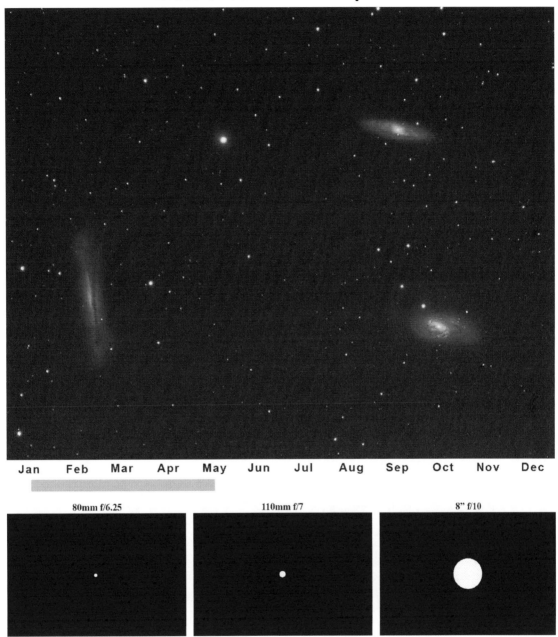

About the target: M65 is at the upper right, M66 at the lower right (both are intermediate spiral galaxies), and NGC 3628 (an unbarred spiral galaxy) is to the left. These three are what is commonly called the Leo Triplet because they sit in the constellation Leo. M65 is an intermediate spiral galaxy approximately 35,000,000 light years away as is NGC3628, however M66 is approximately 36 million light years away.

Where it is:
Right ascension: 11h 20m 15.0s
Declination: +12° 59′ 30″

Imaging the target: Because of my field of view, I always shoot M65 and M66 in the same frame. In fact, I always include NGC 3628 in the field as well. This takes an interesting shot of a small galaxy and turns it into something really special in my opinion. These three are commonly called the Leo Triplet.

These three little galaxies are not only beautiful and show lots of detail, they are also remarkable in that the same exposure can be used on all three. This makes them perfect to sit in the same frame. To extract that much detail from such little objects I used 32 images at 200 seconds at ISO 800 and 25 darks. I would really like to go back and shoot deeper to get a better background field and then merge this image with that one as the background here is not very spectacular. This is high in the sky in winter so thermal noise was not an issue, even with these little targets.

60° Naked-eye Chart. To mag: 6

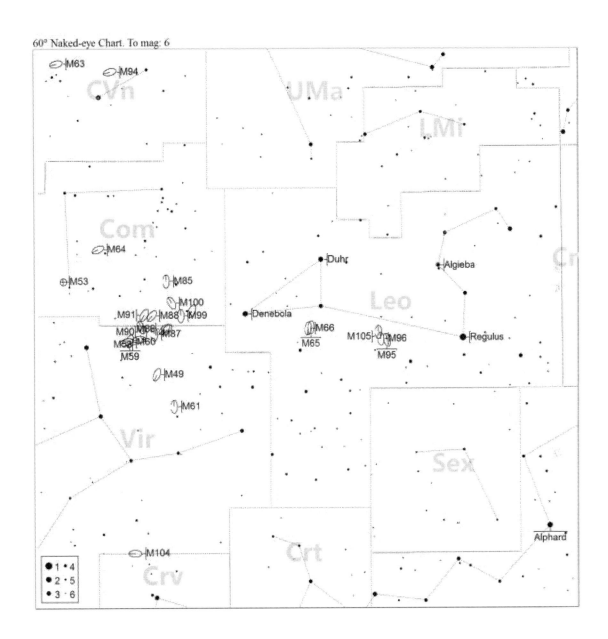

Messier 67:

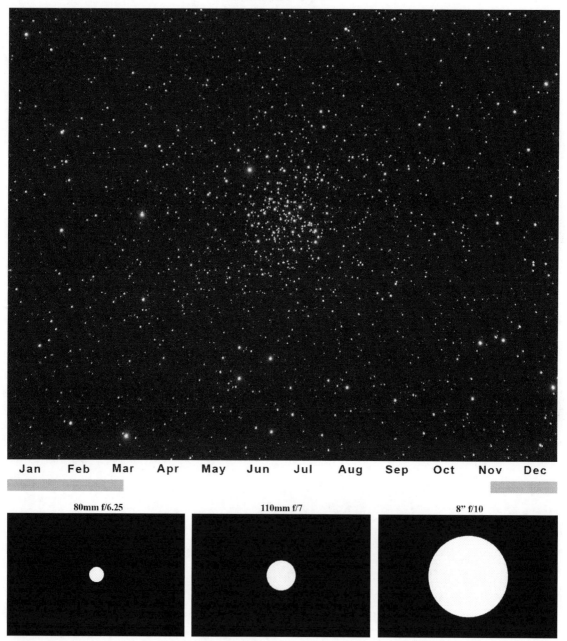

About the target: Discovered by German astronomer Johann Gottfried Koehler in 1779, this open cluster is in the constellation of Cancer and is one of the oldest known open clusters in the sky at somewhere near 4 billion years old. The cluster is approximately 2,700 light years away with a diameter of 20 light years and is estimated to contain somewhere over 500 stars. Messier cataloged this cluster on the 6th of April, 1780.

Where it is:
Right ascension: 08h 51.4m
Declination: +11° 49'

Imaging the target: This is yet another unremarkable open cluster which at least does exhibit a pretty structured dense core so it is relatively easy to find and image. A set of 20 medium exposure 240 second images at ISO 800 should work fine with minimal stretching to control the star colors. In fact, you may want something even shorter such as 150 seconds if you watch your processing. Since this target is high in the winter, thermal noise is not a problem.

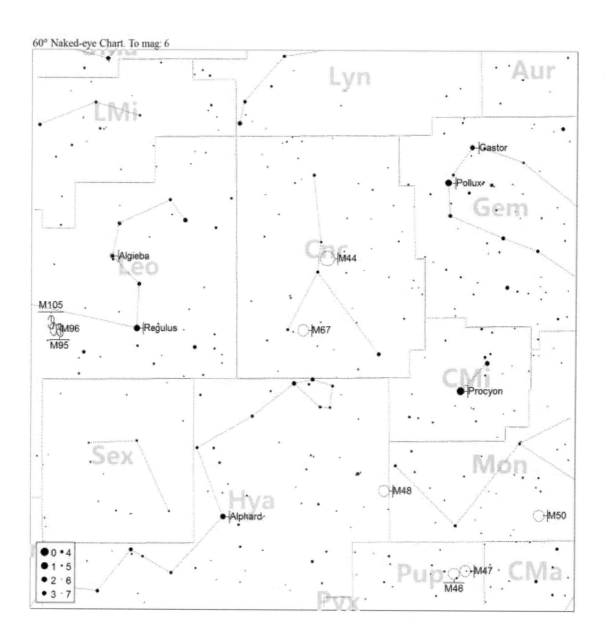

Messier 68:

About the target: Roughly 33,000 light years away in the constellation of Hydra is this globular cluster discovered by Charles Messier on the 9th of April, 1780. The thousands of stars that make up this globular cluster are spread across some 106 light years. Messier said in his logs "it is very faint, very difficult to see with the refractors; near it is star of sixth magnitude."

Where it is:
Right ascension: 12h 39m 27.98s
Declination: −26° 44′ 38.6″

Imaging the target: I like a good globular cluster, but this one really doesn't do it for me. It does have a reasonably structured core but the spread and background field are just a little too pedestrian for me. The good news is that one set of 240 second ISO 800 exposures on my equipment should be more than enough to do a reasonable job capturing this winter target. Do watch out for that bright blue star in the lower right, HIP 61621, as it will blow out very easily.

60° Naked-eye Chart. To mag: 6

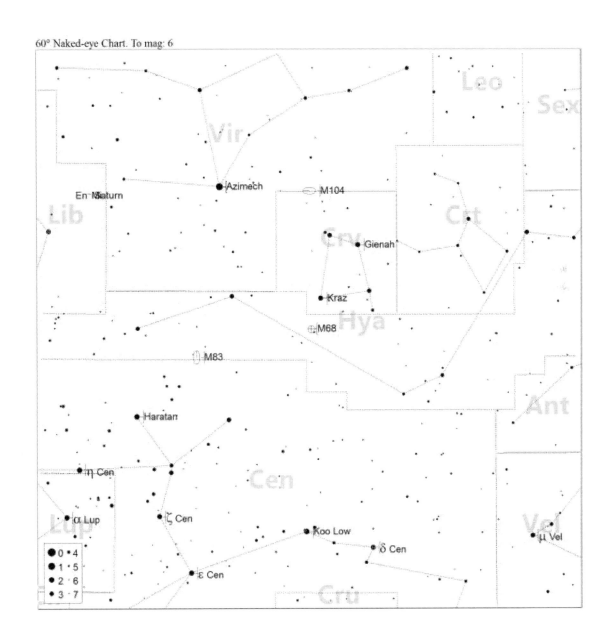

Messier 69:

About the target: The constellation Sagittarius houses this globular cluster that was discovered by Charles Messier on the 31st of August, 1780. The cluster sits some 30,000 light years away and spans 84 light years in diameter, making it fairly small, yet it is quite bright. It is slightly larger and brighter than its very close neighbor, M70, which was discovered by Messier on the same night.

Where it is:
Right ascension: 18h 31m 23.10s
Declination: −32° 20' 53.1"

Imaging the target: This is the little globular that could. In one single set of exposures of 240 seconds at ISO 800 I managed to get a nice structured core, good spread and really nice background field of stars. Very little stretching assured I captured some nice colors for all those blue stars that really stand out against the predominately yellow star field. The only catch here is that you should probably shoot quite a few frames, say 30 or so, as this target is up when the temperatures are high so thermal noise can be a small problem.

60° Naked-eye Chart. To mag: 6

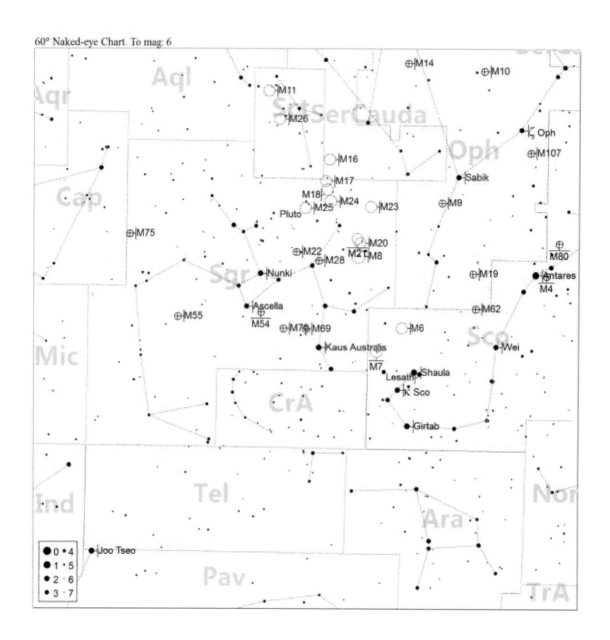

Messier 70:

About the target: The constellation Sagittarius houses this globular cluster that was discovered by Charles Messier on the 31st of August, 1780. The cluster sits some 30,000 light years away and spans 68 light years in diameter, making it fairly small, yet it is quite bright. It is slightly smaller and dimmer than its very close neighbor, M69, which was discovered by Messier on the same night.

Where it is:
Right ascension: 18h 43m 12.76s
Declination: −32° 17′ 31.6″

Imaging the target: Here is another rather disappointing globular cluster. Small, and barely identifiable as a globular, even the nice background field of stars really does little to make this a good image. Additionally, this is high in the sky when the weather is warming up so thermal noise is starting to be a real issue here. I would suggest at least 20 medium exposure images, 240 seconds at ISO 800 for me, and minimize the stretching to keep as much star color as you can.

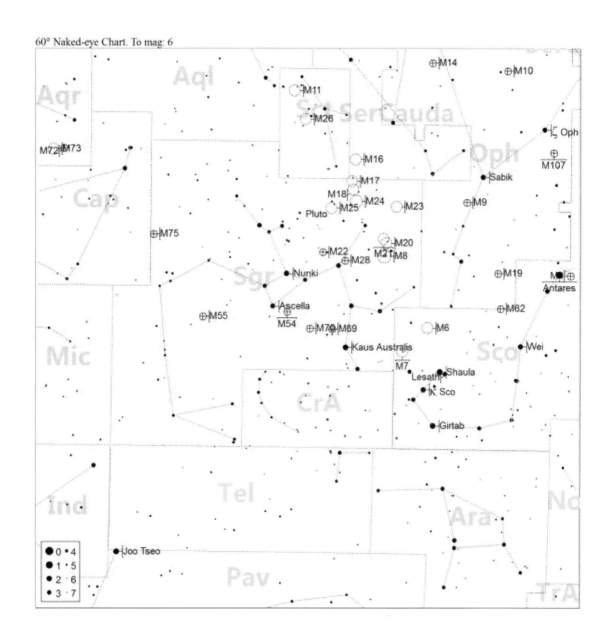

Messier 71:

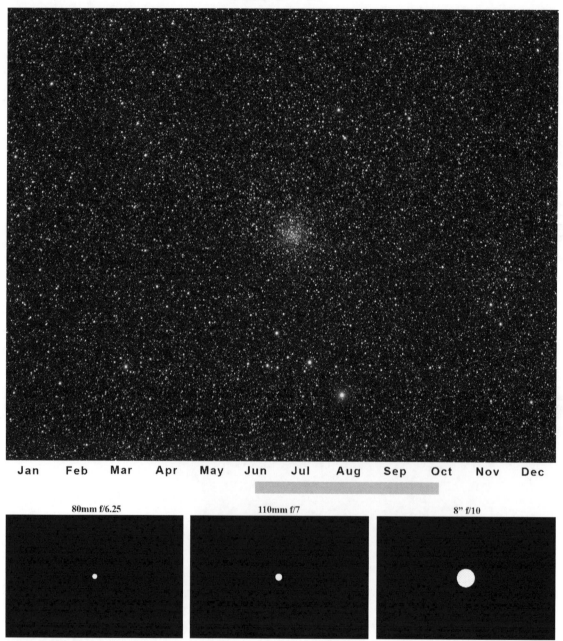

About the target: Swiss astronomer Philippe Loys de Chéseaux discovered this globular cluster in Sagittal around 1746. From about 12,000 light years away and spanning only 26 light years in diameter, this is a small and loosely packed globular. M71 is a quite young globular of around 9 billion years in age, and was long thought to be a fairly dense open cluster.

Where it is:
Right ascension: 19h 53m 46.49s
Declination: +18° 46′ 45.1″

Imaging the target: This so called globular cluster is extremely loose with no real "core" to speak of. This area of the sky however is quite bright so I was able to use fairly short exposures of 150 seconds at ISO 800 to get a very rich background field of stars with some interesting color. I was hoping to get more color although I think that may have been due to a little more stretching than I wanted. More light frames should solve the problem and provide a better set of star colors if you shoot color.

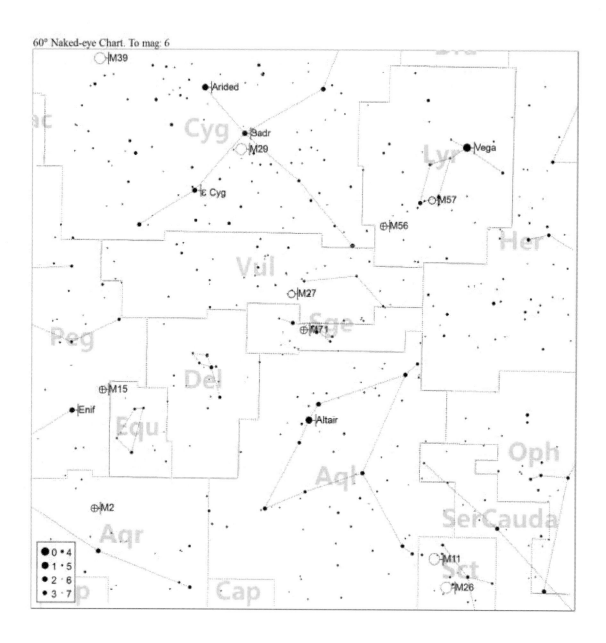

Messier 72:

About the target: On the 29th of August, 1780, French astronomer Pierre Méchain discovered this globular cluster in Aquarius some 55,000 light years away. Being only about 9 billion years old this globular is considered quite young, yet is well defined and includes over 40 known variable stars.

Where it is:
Right ascension: 20h 53m 27.70s
Declination: −12° 32′ 14.3″

Imaging the target: There really isn't a redeeming quality to this globular cluster, no real core, barely any spread and the background field is relatively poor. The Hubble version of this image is pretty impressive however. I shot this at 240 seconds at ISO 800. Pushing a little deeper to 300 seconds or even 400 seconds might get a little better spread and field but be very careful with this approach or the core may blow out. About the only way you could get this to look pretty good is to do two sets of images, one for the core at about 240-300 seconds and one for the spread and field. Possibly even splitting up the field and spread and doing separate exposures for both of them might be an alternative.

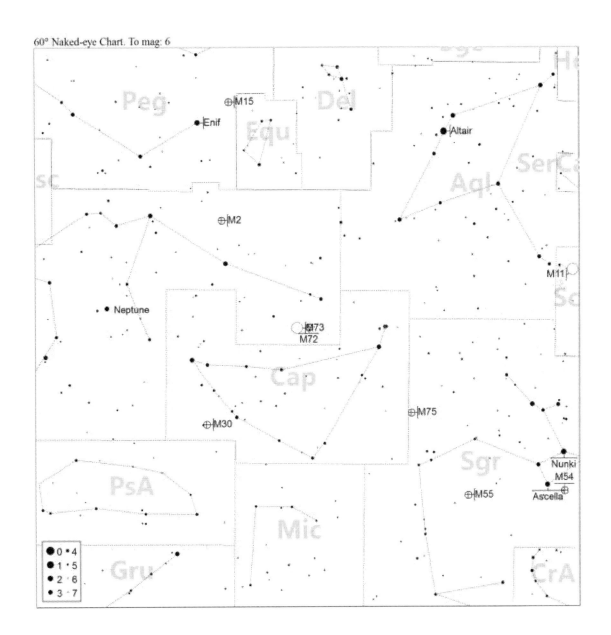

Messier 73:

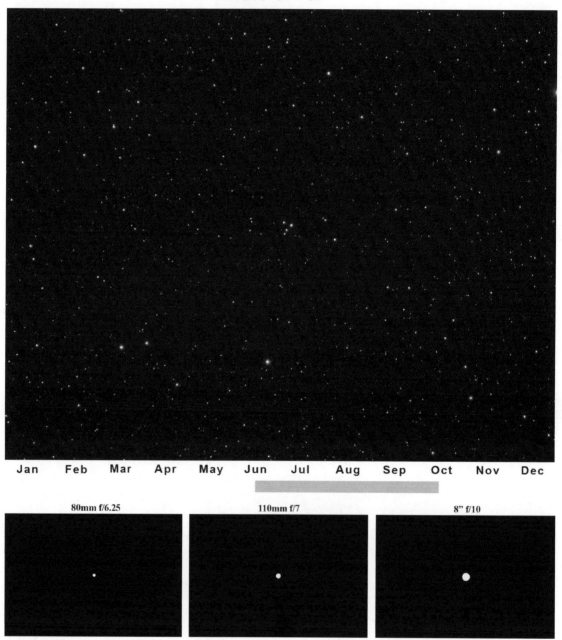

About the target: Charles Messier discovered this asterism (a small visual cluster of stars making a pattern in the sky) in Aquarius on the 4th of October, 1780. While only visually close to each other as viewed from earth, this makes a small but distinguishable pattern that is easy to see and image.

Where it is:
Right ascension: 20h 58m 54s
Declination: −12° 38′

Imaging the target: This asterism is simply four stars in a V pattern with virtually nothing around it so there is not a lot of magic to capturing it. It is however warming up when this target is up in the sky so you might consider taking 20 or more 240 second exposures at ISO 800 for my equipment to make sure the sky shows no thermal noise. No real processing is necessary, even when shooting color. You could however do some minor stretching of the midpoints to bring out a little more of the background stars but be careful not to stretch too far and strip the color from the brighter stars.

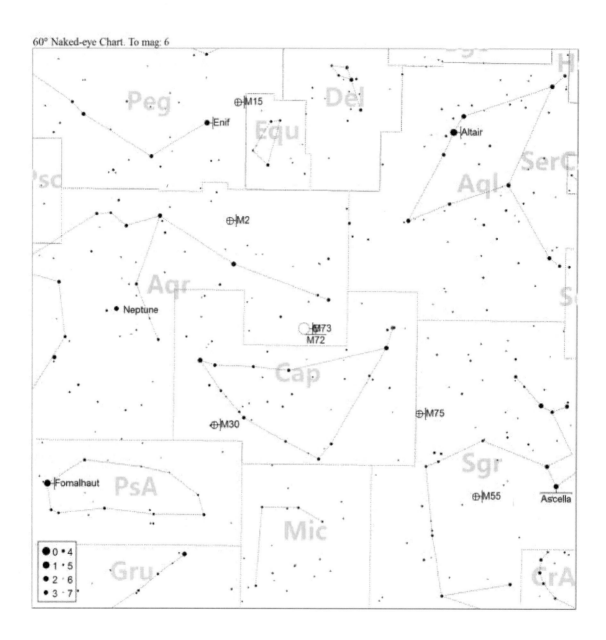

Messier 74:

About the target: Appearing in the constellation Pisces, this spiral galaxy was discovered by French astronomer Pierre Méchain in 1780. Sitting some 32 million light years away it is considered a face-on galaxy in that the flat spiral plane is facing us so we get the best view of its spiral arms. An interesting note is that this galaxy has been home to three recorded supernovae in the 21st century and we could see many more since it is estimated to contain some 100 billion stars.

Where it is:
Right ascension: 01h 36m 41.8s
Declination: +15° 47' 01"

Imaging the target: This nice little spiral galaxy presents a good photo op for a spiral galaxy as it is facing directly towards us and is high in the sky in the fall months. With temperatures cooling off so thermal noise is less of an issue you can shoot deep, 300-480 seconds at ISO 800 for 30 or more frames and really pull some detail out of this target. One initial stretch with levels should be enough to get a good look at the spiral arms, then use curves from there on in small adjustments to really get the detail and colors to pop out. You may want to isolate the galaxy with a mask before doing too much and be sure you isolate it before you preform the final sharpening that will make those arms stand out against the rest.

There is also some nice color stars in the field as well so make sure you are careful about preserving them.

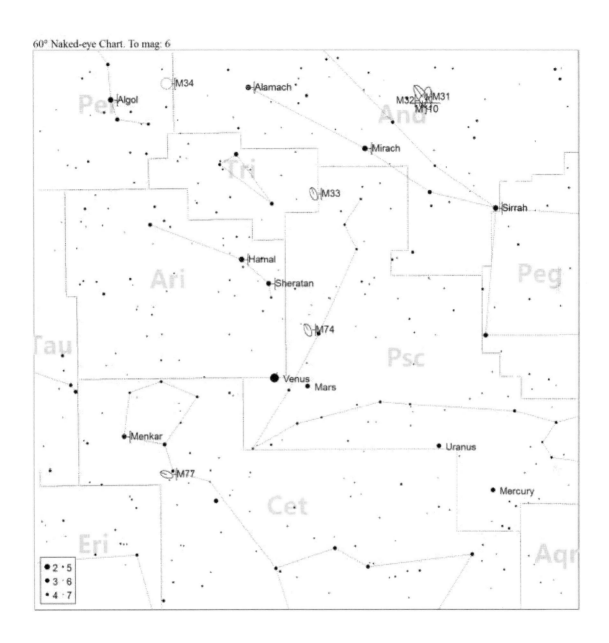

Messier 75:

About the target: This globular cluster in Sagittarius appears as a small, dense globular in images. It spans almost 135 light years and is at a distance of almost 68,000 light years away. Discovered in 1780 by French astronomer Pierre Méchain it is considered to be one of the densest known globular clusters.

Where it is:
Right ascension: 20h 06m 04.75s
Declination: −21° 55′ 16.2″

Imaging the target: Now here is what looks like an interesting tiny globular cluster. In a really small scope it can easily be mistaken for a star. In my scope it is obviously a globular cluster but is too small to catch my eye. In something like an 8" SCT however this could prove to be a very interesting target. Shooting a single set of medium exposure frames, 240 seconds at ISO 800 for me, reveals a pretty nice structure, even at this size. I would love to reshoot this target with a longer scope to see what it can do. In the meantime keep in mind that it is starting to warm up when this target is up so be sure to shoot 20 or more lights to minimize thermal noise and maximize detail.

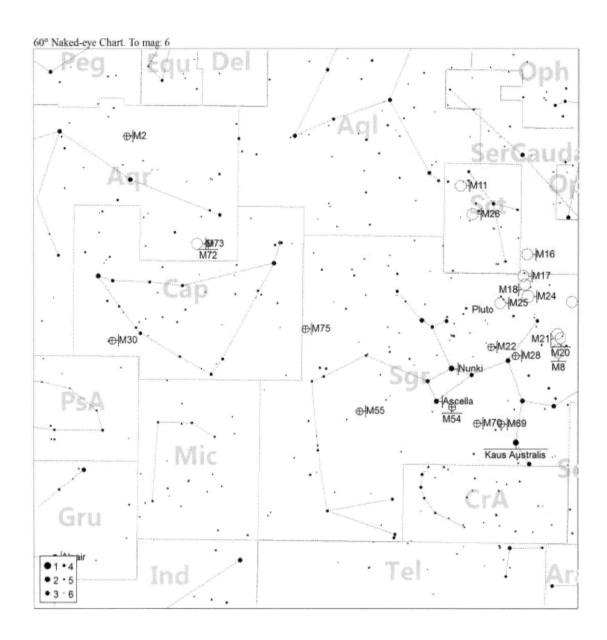

Messier 76: The Little Dumbbell Nebula

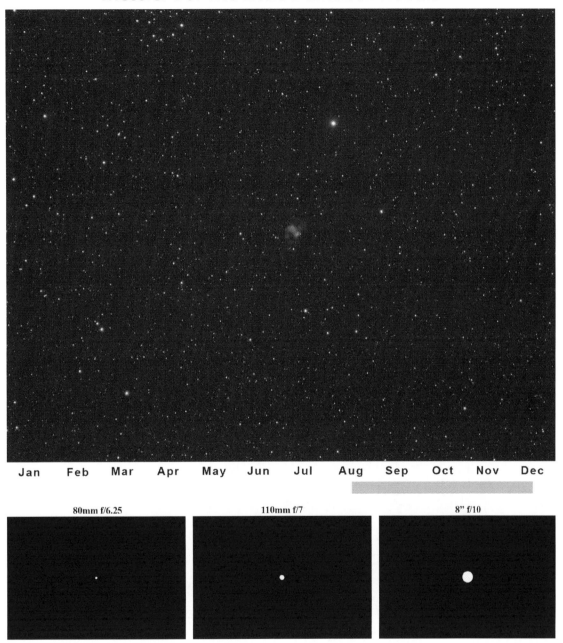

About the target: Discovered in 1780 by French astronomer Pierre Méchain in the constellation Perseus. This planetary nebula (a shell of ionized gas ejected from the surface of a red giant star in its transformation to a white dwarf late in life) is some 2,500 light years away and spans some 1.2 light years in diameter making it a very small and faint object in the sky.

Where it is:
Right ascension: 01h 42.4m
Declination: +51° 34′ 31″

Imaging the target: Here is another target that has me wanting a scope with a longer focal length (although then larger targets would be a problem), the Little Dumbbell Nebula. Like its larger brother M27, this is a predominately blue target which means it is pretty hard to capture. In addition it is just starting to warm up when this target is ready for shooting so we get some thermal noise, and this can be quite a problem with small faint targets.

As you would expect, it responds well to a lot of frames (20+) which helps get the detail while minimizing the noise. A little surprisingly it can work with medium exposures as well; something in the 150-240 second range seems to work pretty good which again helps minimize the noise. There is some nice star color in the background field so careful stretching is required. I achieved fair results using curves but it could have been better with more frames and a mask for stretching.

Messier 77:

About the target: Roughly 47 million light years away in the constellation of Cetus, this barred spiral galaxy was discovered by French astronomer Pierre Méchain on the 29th of October, 1780. Although appearing tiny to us, it is actually one of the largest in this list at over 170,000 light years across.

Where it is:
Right ascension: 2h 42m 40.7s
Declination: −00° 00′ 48″

Imaging the target: This little spiral galaxy is unfortunately one of the many galaxies in the Messier list that just do not present enough to get much detail out of it. You can however make an interesting field if you shoot a lot (40-50) of fairly deep (300-480 seconds at ISO 800) frames as there are other galaxies in the area that should pop out without too much fuss. You will need a lot of frames because some of the galaxies are pretty faint and of course because this target is up in the middle of the summer so thermal noise is a killer. In fact, looking at the image I included for this target, you can see how I failed to take enough frames to deal with the thermal noise and it really did a number on both the galaxies shown.

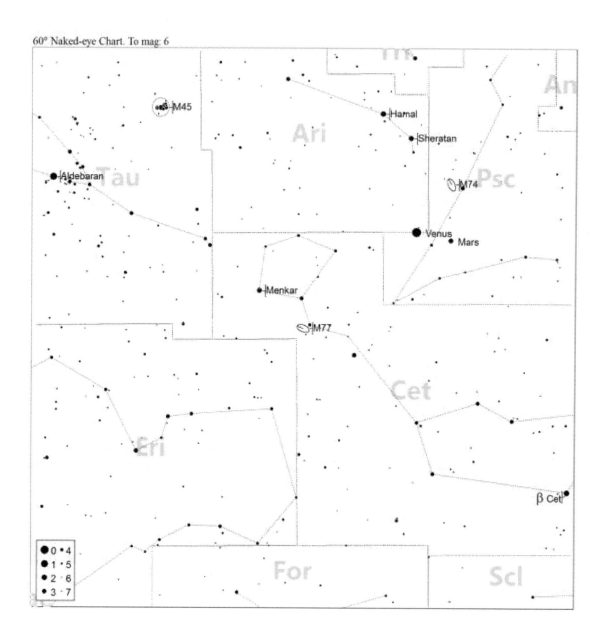

Messier 78:

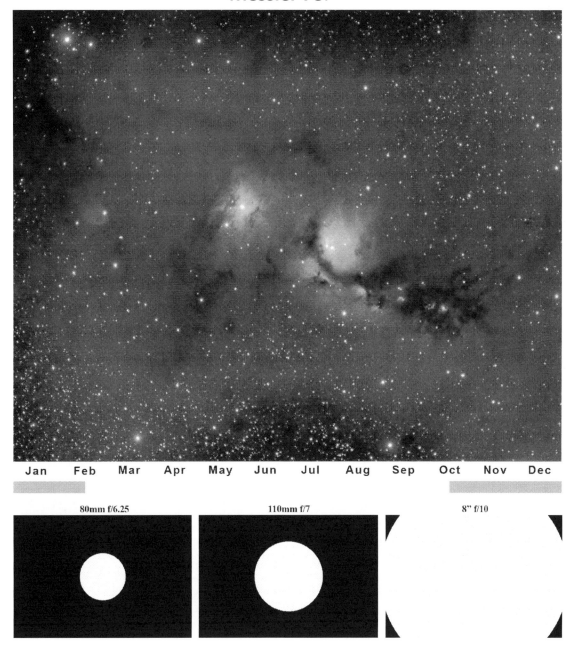

| Jan | Feb | Mar | Apr | May | Jun | Jul | Aug | Sep | Oct | Nov | Dec |

80mm f/6.25   110mm f/7   8" f/10

About the target: Part of the Orion Molecular Cloud Complex in the constellation of Orion sits the brightest reflection nebula of that complex, M78. Discovered in 1780 by French astronomer Pierre Méchain it sits some 1,600 light years from earth and spans some 10 light years in diameter. Two stars (HD 38563A & HD 38563B) reflect their light off the clouds to cause the hazy patches most amateur astronomers see in their telescopes.

Where it is:
Right ascension: 05h 46.7m
Declination: +00° 03′

Imaging the target: Finally, my little pet target. This target can change your life. Ok, maybe not, but it certainly can make you stand up and take notice. The first time you shoot this target you will probably think I am insane, all you will see is two tiny fuzzy spots of nebulosity and some stars. Less than thrilling. Keep going.

Getting a good result from this target requires a lot of long exposures. In the example image here there are 70 (yes, seventy) 300 second exposures at ISO 800 (just under 6 hours of data) and it needs a lot more. My goal is to reshoot this target this year with somewhere near 100 480 second exposures (just over 13 hours of data) to see what I can get out of it. Stretching this target is a lot of work. The initial stretch is using levels with all the remaining stretches being with curves in separate layers to make it easy to go back and restretch when I mess up. I probably have ten or more hours in processing this image after it was shot.

Now you may think that since this target is up in the fall that thermal noise would not be an issue, you would be wrong. The detail is so faint and requires so much stretching that all noise, including thermal, is a serious challenge. This is one reason for the high number of frames. It is also why I would recommend scheduling your time to shoot this target into several consecutive nights with reasonably consistent temperatures.

The results however, are worth it. If you want a serious challenge with excellent potential, this is your target.

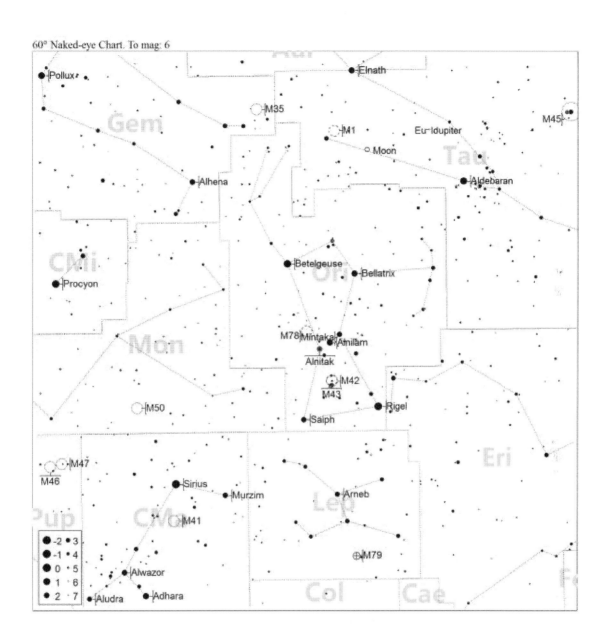

Messier 79:

About the target: This globular cluster is one of two extragalactic (from outside our galaxy) globular clusters on the Messier list and is believed to be from the Canis Major Dwarf Galaxy some 41,000 light years away. Sitting in the constellation Lepus, it was discovered in 1780 by Pierre Méchain. This cluster is thought to be some 12 billion years old and was added by Messier to his list on the 17th of December, 1780.

Where it is:
Right ascension: 05h 24m 10.59s
Declination: −24° 31' 27.3"

Imaging the target: This is another little globular cluster that is difficult to turn into something interesting to image. At relatively short exposures the core tends to blow out as it is very compact. The spread and field however are pretty faint and will require a much longer exposure to bring anything reasonable out. I would suggest you shoot the core at a short exposure not to exceed 150 seconds at ISO 800 (with my equipment), and then shoot a separate set of pretty deep exposures around 240-300 seconds at ISO 800. Combine the two sets of images in a HDR program or using layers to form one image.

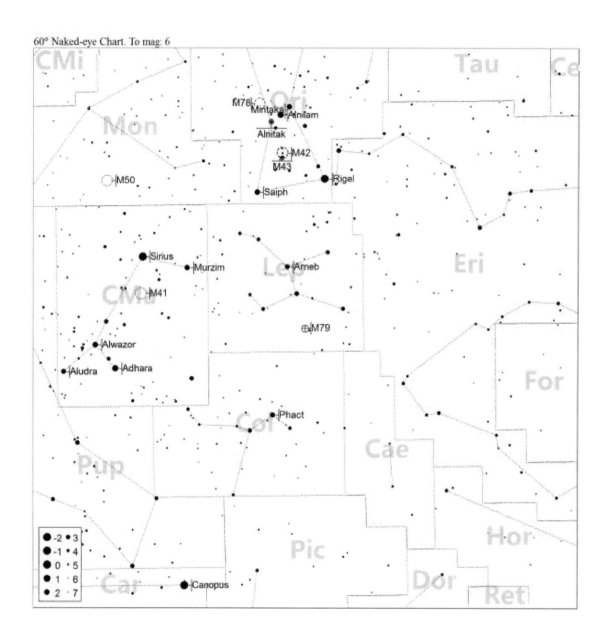

Messier 80:

About the target: This globular cluster in Scorpius was discovered on the 4th of January, 1781, by Charles Messier. Approximately 32,000 light years away and less than 100 light years in diameter, this appears as a fairly small but dense cluster. In 1860 there was a nova reported by German astronomer Georg Friedrich Julius Arthur von Auwers and a second was recorded on film in 1938.

Where it is:
Right ascension: 16h 17m 02.41s
Declination: −22° 58′ 33.9″

Imaging the target: Just like M79, this is another little globular cluster that is difficult to turn into something interesting to image. At relatively short exposures the core tends to blow out as it is very compact. The spread and field however are pretty faint and will require a much longer exposure to bring anything reasonable out. I would suggest you shoot the core at a short exposure not to exceed 150 seconds at ISO 800 (with my equipment), and then shoot a separate set of pretty deep exposures around 240-300 seconds at ISO 800. Combine the two sets of images in a HDR program or using layers to form one image.

60° Naked-eye Chart. To mag: 6

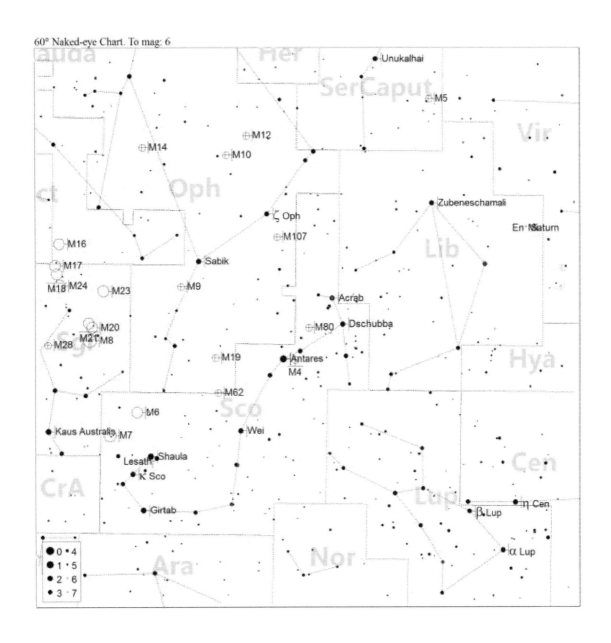

Messier 81: Bode's Galaxy

About the target: M81 is on the right, M82 on the left. M81 is a spiral galaxy about 12 million light years away in the constellation Ursa Major. It was named after its discoverer, German astronomer Johan Bode in 1774. M82 is named the cigar galaxy, well, because it looks like a cigar! It is roughly the same distance as M81 and is a starburst galaxy.

Where it is:
Right ascension: 09h 55m 33.2s
Declination: +69° 3' 55"

Imaging the target: I always shoot M81 and M82 as a pair, and if they will fit together in your field of view I recommend you do the same. Much like M65 and M66 these two galaxies respond excellently to a single set of exposures and to being stretched together. To get the golden brown lines of detail on M82 and the nice spiral arms on M81 you will need some fairly deep exposures (about 300 seconds at ISO 800 for me) and quite a few frames (I went for 31) to be able to stretch them sufficiently. Thermal noise is not too much of an issue as this target is up in winter so that helps a little. You will need to use curves primarily to get the outer arms on M81 to really pop without blowing out the central core, or blowing out M82 so much those golden bands are no longer visible.

On several other galaxies I have suggested some sharpening but here I have had less than optimal results. You might try a little sharpening but I recommend you do it in very small amounts and spend some time looking at the results before you commit.

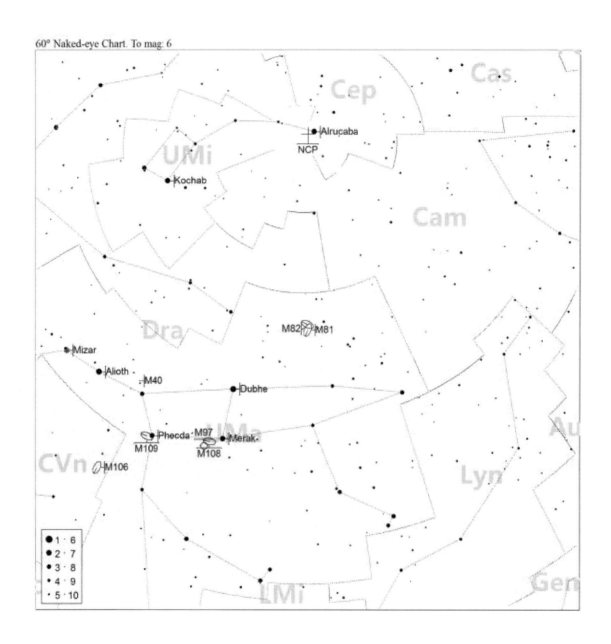

Messier 82: The Cigar Galaxy

About the target: M81 is on the right, M82 on the left. M81 is a spiral galaxy about 12 million light years away in the constellation Ursa Major. It was named after its discoverer, German astronomer Johan Bode in 1774. M82 is named the cigar galaxy, well, because it looks like a cigar! It is roughly the same distance as M81 and is a starburst galaxy.

Where it is:
Right ascension: 09h 55m 52.2s
Declination: +69° 40′ 47″

Imaging the target: I always shoot M81 and M82 as a pair, and if they will fit together in your field of view I recommend you do the same. Much like M65 and M66 these two galaxies respond excellently to a single set of exposures and to being stretched together. To get the golden brown lines of detail on M82 and the nice spiral arms on M81 you will need some fairly deep exposures (about 300 seconds at ISO 800 for me) and quite a few frames (I went for 31) to be able to stretch them sufficiently. Thermal noise is not too much of an issue as this target is up in winter so that helps a little. You will need to use curves primarily to get the outer arms on M81 to really pop without blowing out the central core, or blowing out M82 so much those golden bands are no longer visible.

On several other galaxies I have suggested some sharpening but here I have had less than optimal results. You might try a little sharpening but I recommend you do it in very small amounts and spend some time looking at the results before you commit.

60° Naked-eye Chart. To mag: 6

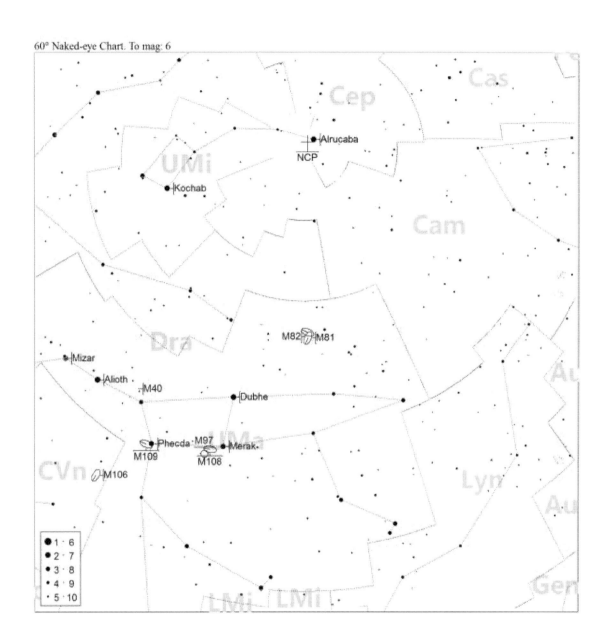

Messier 83: The Southern Pinwheel Galaxy

About the target: This barred spiral galaxy is roughly 15 million light years away in the constellation Hydra and is one of the closest of its kind to us, making it one of only a few galaxies visible in handheld binoculars. It is unique in that there have been six supernovae observed in this galaxy with the latest being SN 1983N. Discovered by French astronomer Nicolas Louis de Lacaille on the 23rd of February, 1752, it was not until the 17th of February 1781 that it was added to Messier's list.

Where it is:
Right ascension: 13h 37m 00.9s
Declination: −29° 51′ 57″

Imaging the target: I have yet to get a shot of M83 that I actually like. This object is my curse. Every time I go to shoot it, something goes horribly wrong. Don't worry, you will get one of those targets too :-)

The Southern Pinwheel is a pretty faint galaxy with some nice detail, but you will have to shoot a lot of very deep images to pull things out. I would suggest starting at 480 second exposures at ISO 800 for equipment similar to mine, and shoot at least 30-40 shots. Stretching is fairly straight forward for a galaxy; start with a little bit of levels to get it to come to life and then use small tweaks in curves to get all the arm detail to really pop. Once that is done, isolate the galaxy with a mask and apply some sharpening to get the details to stand out better. Be careful with the sharpening though as it is very easy to overdo.

You should not have a problem with thermal noise as this target is up in the winter as long as you shoot it early in the morning although you will have some serious noise from the length of exposures.

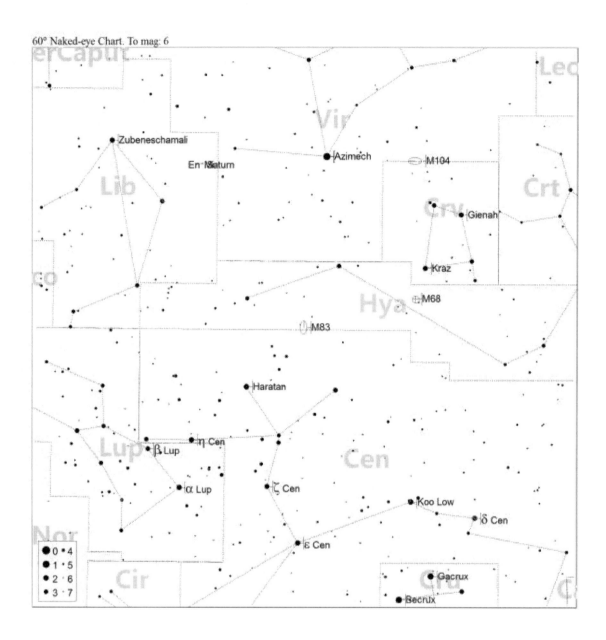

Messier 84:

Jan Feb Mar Apr May Jun Jul Aug Sep Oct Nov Dec

80mm f/6.25　　　　110mm f/7　　　　8" f/10

About the target: "the center it is a bit brilliant, surrounded with a slight nebulosity" was part of the description Charles Messier wrote on the 18th of March, 1781, regarding this lenticular galaxy in Virgo, part of the Virgo SuperCluster of galaxies. Even at 60 million light years distance this is one of the brighter galaxies in this part of the sky.

Where it is:
Right ascension: 12h 25m 03.7s
Declination: +12° 53' 13"

Imaging the target: This is a good news, bad news situation. The bad news is that even the Hubble space telescope hasn't pulled any decent detail out of this fuzzy blob so you won't either. The good news is that if you look back at the section of the book talking about the Virgo SuperCluster you will see that really cool picture of "the chain"; look at the bottom left and notice that M84 is right there at the start of the chain. This means that even though you will never get a great picture of just M84, you can get an awesome image with M84 in it.

To get something like my image of the chain, shoot deep, 300-480 seconds at ISO 800 and capture at least 30 frames. With targets this small, curves are difficult at best so most of your stretching will be careful use of levels, a little at a time, until you get the desired results.

This is also one of the images that seems more impressive when you label all the galaxies in the image.

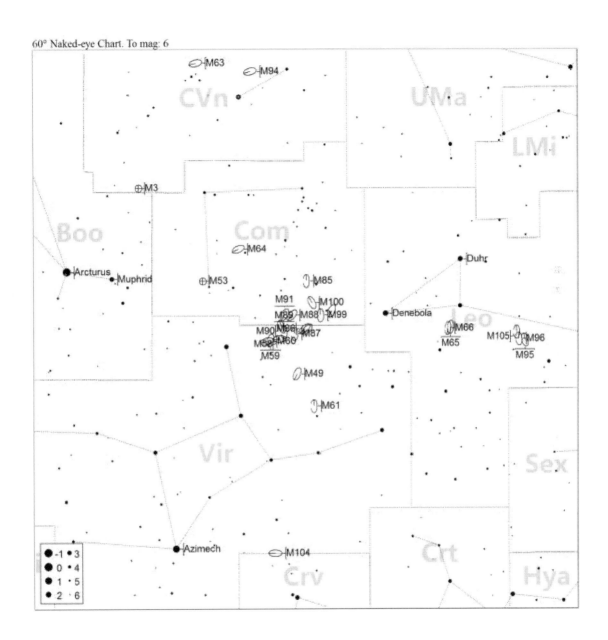

Messier 85:

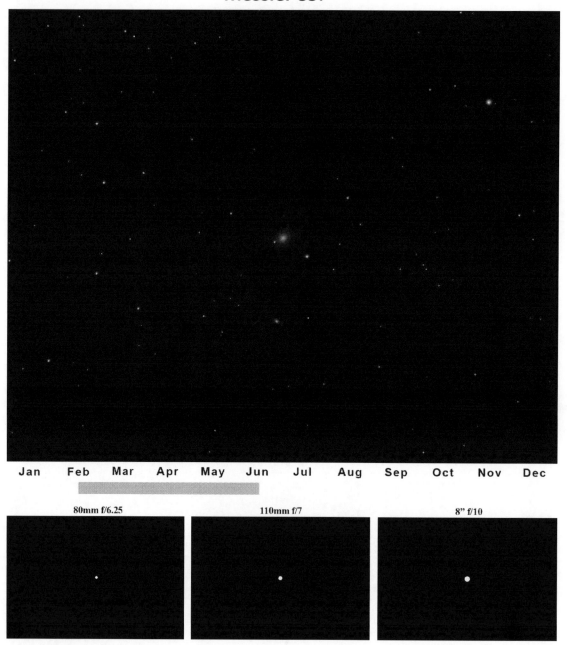

Jan   Feb   Mar   Apr   May   Jun   Jul   Aug   Sep   Oct   Nov   Dec

80mm f/6.25   110mm f/7   8" f/10

About the target: Confusion surrounds this galaxy as the debate continues on whether this is a lenticular or elliptical galaxy sitting some 60 million light years away on the very outer edge of the Virgo SuperCluster. Making this object even stranger, it was where the first luminous red nova (suspected collision of two stars) was identified on the 7th of January, 2006. When Pierre Méchain discovered it in 1781 he probably never expected it to be this interesting.

Where it is:
Right ascension: 12h 25m 24.0s
Declination: +18° 11′ 28″

Imaging the target: Here is another fairly boring lenticular galaxy. The saving feature of this image is that looking closely, you can see a couple of other galaxies as well. Shooting deep, for me somewhere near 480 seconds at ISO 800 and stretching carefully can make some of these galaxies jump out of this image. In my opinion, the best way to shoot these types of targets is when you can get as many targets in one frame as possible such as the picture I included in the front of this book in the section on the Virgo SuperCluster.

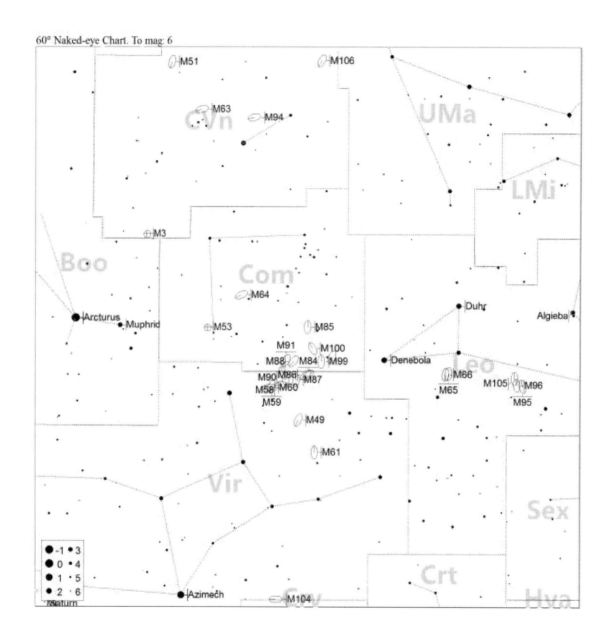

Messier 86:

About the target: At the center of the Virgo SuperCluster sits this elliptical or lenticular galaxy some 52 million light years distant. Discovered by Charles Messier on the 18th of March, 1781, the same night he discovered seven other galaxies, it is surrounded by objects. This galaxy is known for its blue shift (color spectrum shift indicating movement towards or away from us) as it is the Messier object moving towards us at the greatest speed, 419 km/s.

Where it is:
Right ascension: 12h 26m 11.7s
Declination: +12° 56′ 46″

Imaging the target: This is pretty much a duplicate of M84 and is also a good news, bad news situation. The bad news is that even the Hubble space telescope hasn't pulled any decent detail out of this fuzzy blob so you won't either. The good news is that if you look back at the section of the book talking about the Virgo SuperCluster, you will see that really cool picture of "the chain"; look at the bottom left and notice that M86 is near the start of the chain. This means that even though you will never get a great picture of just M86, you can get an awesome image with M86 in it.

To get something like my image of the chain, shoot deep, 300-480 seconds at ISO 800 and capture at least 30 frames. With targets this small curves are difficult at best so most of your stretching will be careful use of levels, a little at a time, until you get the desired results.

This is also one of the images that seems more impressive when you label all the galaxies in the image.

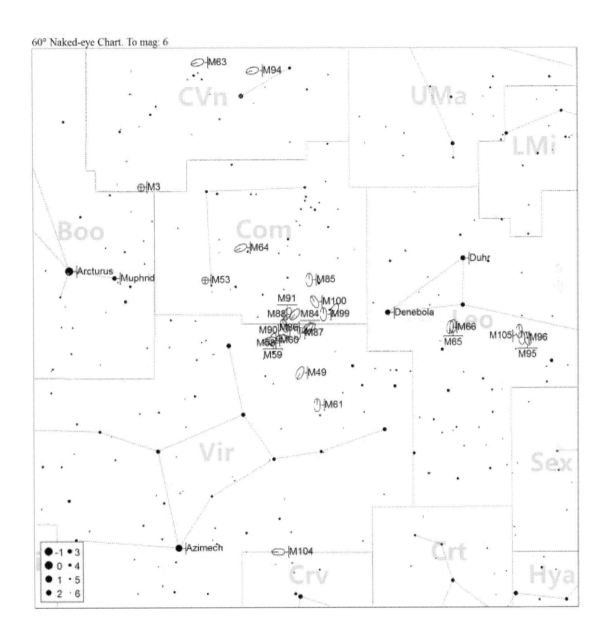

Messier 87:

About the target: This supergiant elliptical galaxy appears small because it is roughly 53 million light years away and is at the very heart of the Virgo SuperCluster of galaxies, in the constellation of Virgo. This one image contains probably over twenty visible galaxies including NGC4478, NGC4476, IC3443 and many more. Discovered by Messier on the 18th of March, 1781, and despite its distance, it is one of the brightest galaxies in the SuperCluster. An interesting item of note is that this galaxy has a supermassive black hole that is ejecting a stream of matter from the center of the galaxy out some 5,000 light years.

Where it is:
Right ascension: 12h 30m 49.42338s
Declination: +12° 23' 28.0439"

Imaging the target: Here is another fairly boring elliptical galaxy. The saving feature of this image is that looking closely; you can see several other galaxies as well. Shooting deep, for me somewhere near 480 seconds at ISO 800 and stretching carefully can make some of these galaxies jump out of this image. In my opinion, the best way to shoot these types of targets is when you can get as many targets in one frame as possible such as the picture I included in the front of this book in the section on the Virgo SuperCluster. There is probably somewhere near 50 galaxies in my field of view of this image but I did not shoot deep enough to really bring them out.

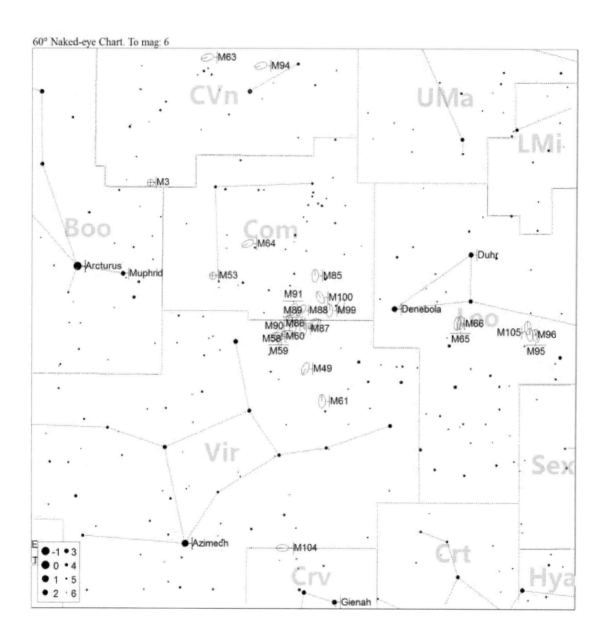

Messier 88:

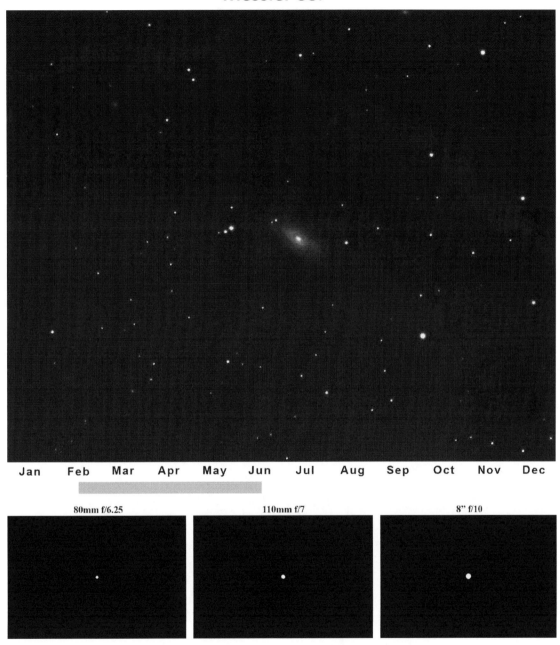

About the target: Discovered by Charles Messier on the 18th of March, 1781, the same night that he discovered seven other galaxies, this spiral galaxy in the Virgo SuperCluster is some 50 million light years away in the constellation Coma Berenices. This galaxy was home to supernova 1999cl and was one of the first galaxies to be identified as a spiral galaxy. It is thought to stretch some 130,000 light years in diameter.

Where it is:
Right ascension: 12h 31m 59.2s
Declination: +14° 25′ 14″

Imaging the target: Now here is a tough little spiral galaxy for you. Presenting at the start of summer, you will be fighting not only a lot of detail in a very small target, but also a lot of thermal noise. This target would probably present a better image in a long focal length scope like an 8" SCT or better, but even in my short focal length refractor you can start to get some detail with enough images (I suggest 30+) going deep enough (I suggest 300-480 seconds). Stretching is also going to be difficult with all the thermal noise so remember that the more images you shoot, the better your result.

An interesting note: this is one of the few spiral galaxies that present this angle (64 degrees) that you can actually get detail out of so it could be worth a little time and effort.

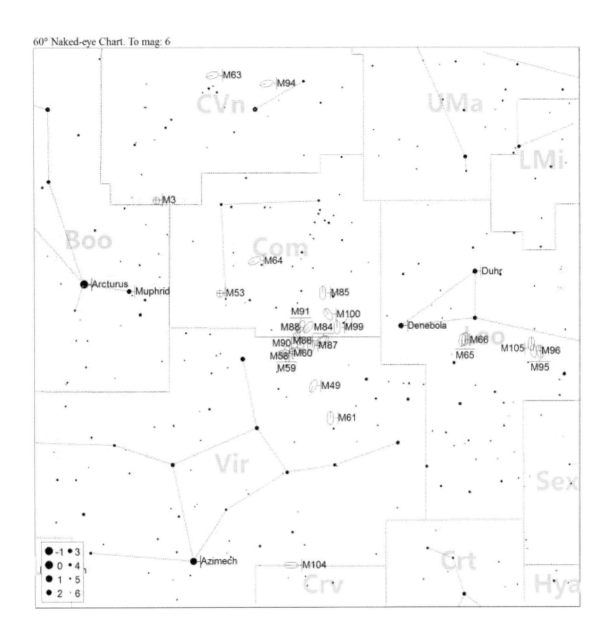

Messier 89:

About the target: This is a nearly perfectly spherical elliptical galaxy, at least to us viewing it from Earth. Located in the Virgo SuperCluster of galaxies in the constellation of Virgo, it was discovered by Charles Messier on March 18th, 1781. At one time this galaxy may have been home to a quasar as it has streams of matter extending approximately 100 thousand light years from the core. Making this galaxy even more scientifically interesting, it is believed to be home to some 2,000 or more globular clusters.

Where it is:
Right ascension: 12h 35m 39.8s
Declination: +12° 33′ 23″

Imaging the target: Here is yet another fairly boring elliptical galaxy. The saving feature of this image again is that looking closely; you can see several other galaxies as well. Shooting deep, for me somewhere near 480 seconds at ISO 800 and stretching carefully can make some of these galaxies jump out of this image. In my opinion, the best way to shoot these types of targets is when you can get as many targets in one frame as possible such as the picture I included in the front of this book in the section on the Virgo SuperCluster. There is probably somewhere near 30 galaxies in my field of view of this image but I did not shoot deep enough to really bring them out. There are however three other galaxies in the image clearly visible.

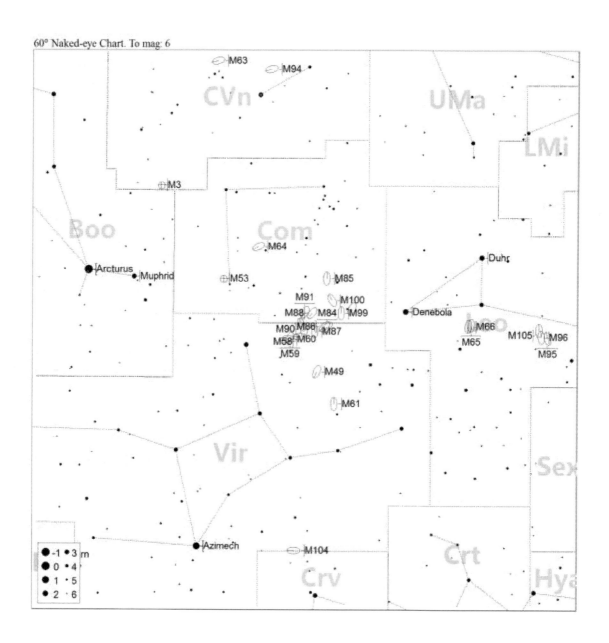

Messier 90:

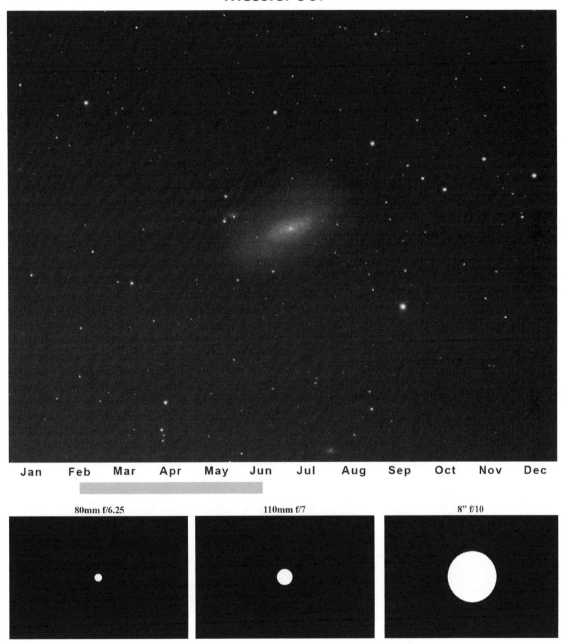

About the target: This spiral galaxy in the Virgo SuperCluster was originally discovered by Charles Messier on the 18th of March, 1781, of which it was one of nine discoveries on that day for Messier. Resting some 60 million light years away, it is the second fastest approaching galaxy, headed towards us at some 383 km/sec.

Where it is:
Right ascension: 12h 36m 49.8s
Declination: +13° 09' 46"

Imaging the target: Much like M88 this is an angled spiral galaxy which presents a good level of detail. This one however has the advantage of being up at the start of spring so thermal noise will be less of an issue. The image you see is 20 frames of 300 seconds at ISO 800 and could really use three to four times as much time to really bring out the details. If you look close you can see the grainy texture of the outer arms that resulted from stretching it way beyond what I should have for the amount of time I had on this target. You can also see that had I put more time on the target, I could have gotten more detail as the arms continue to go out beyond the spiral pretty far.

60° Naked-eye Chart. To mag: 6

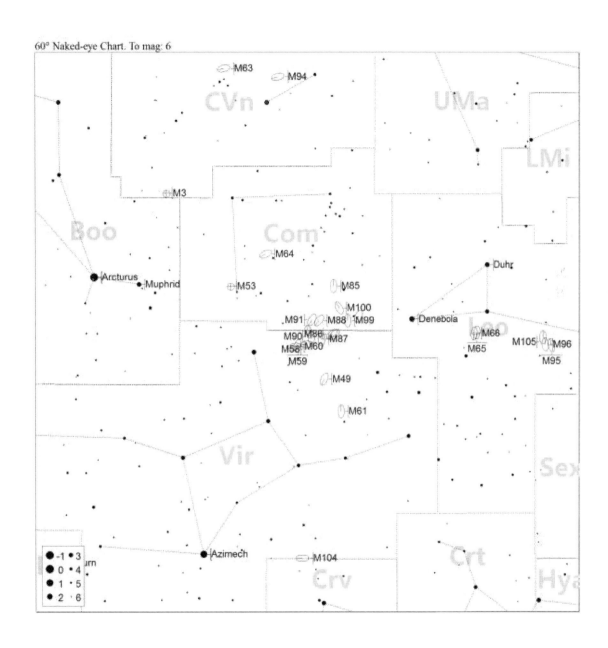

Messier 91:

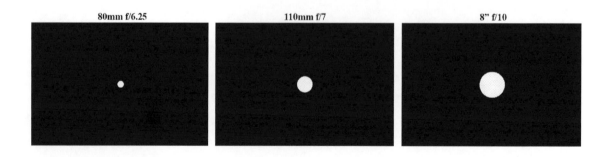

About the target: The constellation Coma Berenices is home to this barred spiral galaxy discovered on the 18th of March, 1781 by Charles Messier. Approximately 63 million miles distant, M91 was originally accidentally left off of Meisser's list but in a 1969 letter to Sky & Telescope magazine, American amateur astronomer William C. Williams demonstrated that M91 was in fact NGC 4548.

Where it is:
Right ascension: 12h 35m 26.4s
Declination: +14° 29′ 47″

Imaging the target: This barred spiral galaxy is much like M90 and M88 in its angle and our ability to capture it. One drawback here is that it is a little later in rising than either of the others which means thermal noise is a bit more of an issue. I would recommend that you start off with at least 30 images of 300-480 seconds at ISO 800 on equipment similar to mine and do your stretching in very small amounts at a time. This galaxy will be a little more difficult to coax out detail from than either M88 or M90 so take your time. I also doubt that you will want to use any kind of sharpening on this particular target unless you have a longer focal length scope such as an 8" SCT as it is just too small.

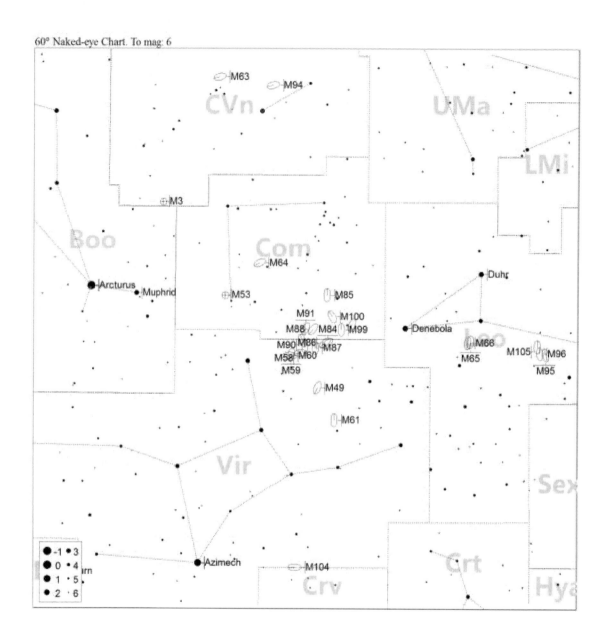

Messier 92:

About the target: Discovered by German astronomer Johann Elert Bode in 1777, this globular cluster in the constellation of Hercules is approximately 27,000 light years from us and has an estimated diameter of 109 light years. Sitting right next to M13, this is one of the brightest clusters in the sky and is approximately 14 billion years old. Messier added this to his catalog on the 18th of March, 1781. Under excellent conditions this object can be seen with the naked eye and using binoculars or better, one can view it under most sky conditions.

Where it is:
Right ascension: 17h 17m 07.39s
Declination: +43° 08′ 09.4″

Imaging the target: This globular cluster really needs multiple exposures to balance out as the core is pretty tightly packed and bright whereas the background field is relatively sparse. You might even want to do three different exposures allowing one just for the spread as well although that may not be necessary. A good starting point might be 20 or more at 150 seconds for the core, 20 or more at 300 seconds for the spread and 30 or more for the background field. The reason to increase the exposures for the background is that this target is high in the middle of summer and thermal noise is really going to become an issue in the faint stars and black of space, especially at those long exposures. Keep any stretching to a minimum to preserve star colors if you are shooting in color.

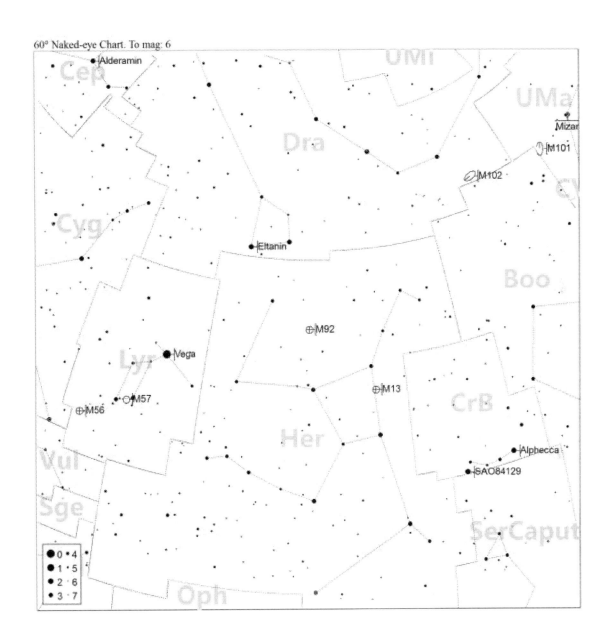

Messier 93:

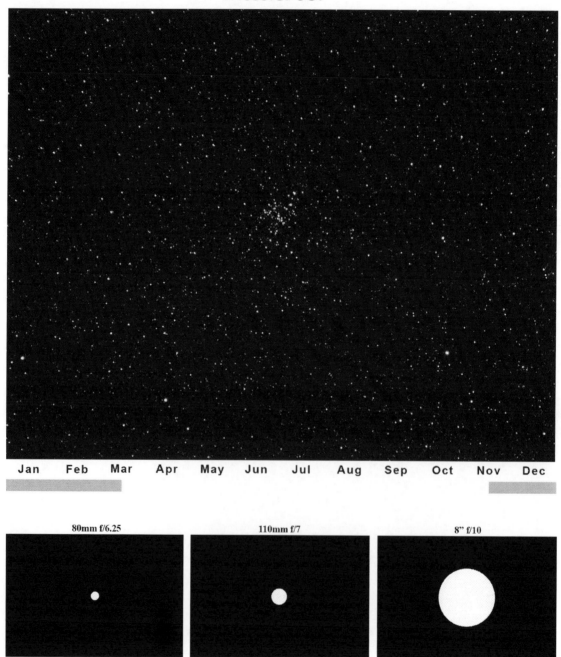

About the target: This 100 million year old open cluster is about 3,600 light years away in the constellation Puppis and was discovered by Charles Messier on the 20th of March, 1781. It appears to house some 80 stars in its 20 or more light year diameter. It seems that even Messier was less than impressed as his notes read "Cluster of small stars, without nebulosity...".

Where it is:
Right ascension: 07h 44.6m
Declination: −23° 52′

Imaging the target: This rather bland open cluster has a fairly compact center that contains two bright yellow stars which can contrast well with the rest of the cluster if you shoot in color. Exposures can be fairly short to medium, 150-240 seconds at ISO 800 for me. You don't need a lot of exposures as thermal noise should be minimal considering the short exposures and that this object is high in the winter months.

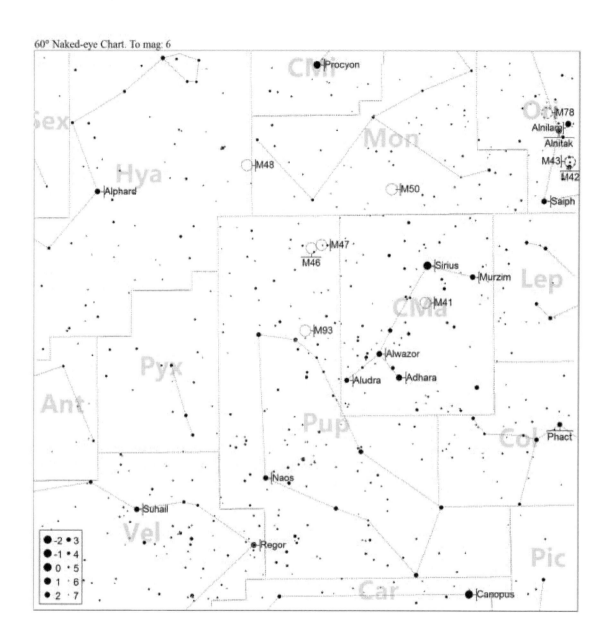

Messier 94:

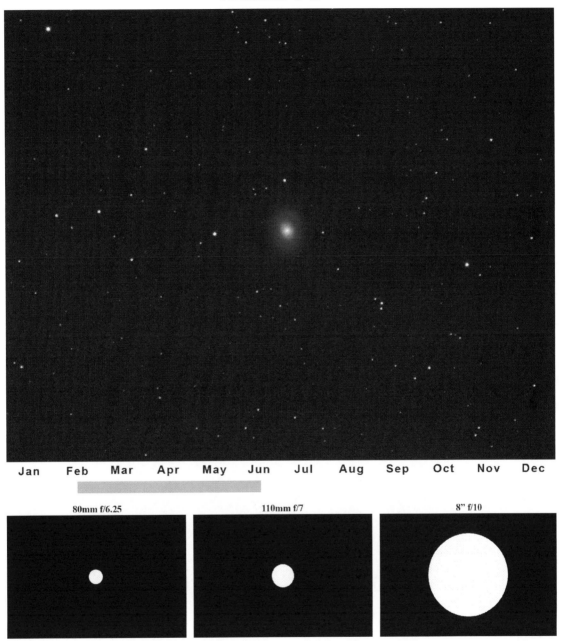

About the target: Discovered by French astronomer Pierre Méchain in 1781, this spiral galaxy appears to contain both an inner and outer ring. The outer ring actually is a set of spiral arms when viewed in IR and UV light wavelengths. Messier included this object in his catalog on the 24th of March, 1781. M94 is estimated to be some 16 million light years away.

Where it is:
Right ascension: 12h 50m 53.1s
Declination: +41° 07′ 14″

Imaging the target: Here we have a near face on spiral galaxy that is rather small in my field of view. Although you can do something with it at medium focal ratios, it would really be better in a long focal length scope such as an 8" SCT. It is bright enough that 150-240 second exposures at ISO 800 were enough for me although I could tell that I could have gone deeper to catch the outer arms that are just barely visible in the image provided. Thermal noise was not really an issue as it was shot in late winter. It looks like some interesting detail is available all the way to the core so I will have to reshoot this target soon.

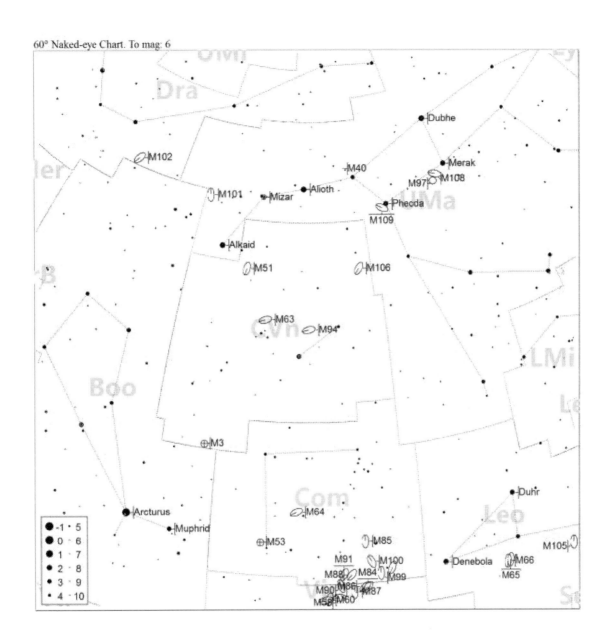

Messier 95:

Jan Feb Mar Apr May Jun Jul Aug Sep Oct Nov Dec

80mm f/6.25   110mm f/7   8" f/10

About the target: Pierre Méchain discovered this barred spiral galaxy in the constellation Leo in 1781. This galaxy sits some 38 million light years away and was cataloged by Messier on the 24th of April, 1781. Recently M95 has been in the news as it was home to SN 2012aw, a supernova in one of the outer arms. M95 was one of eighteen galaxies used in the HST Key Project to help improve the accuracy of astronomical distance measurements.

Where it is:
Right ascension: 10h 43m 57.7s
Declination: +11° 42' 14"

Imaging the target: The barred spiral galaxy here has some real potential even for medium focal length scopes although you will have to go really deep to get them, something in the neighborhood of 30 or more at 480 seconds at ISO 800, or even more and longer. Since this is a winter target, thermal noise is not really a problem but you will still fight the noise of the really long exposures and really dim galaxy. This target definitely is on my list of targets for a reshoot to see what I can make of it as last time I shot too few shots and stretched it too hard.

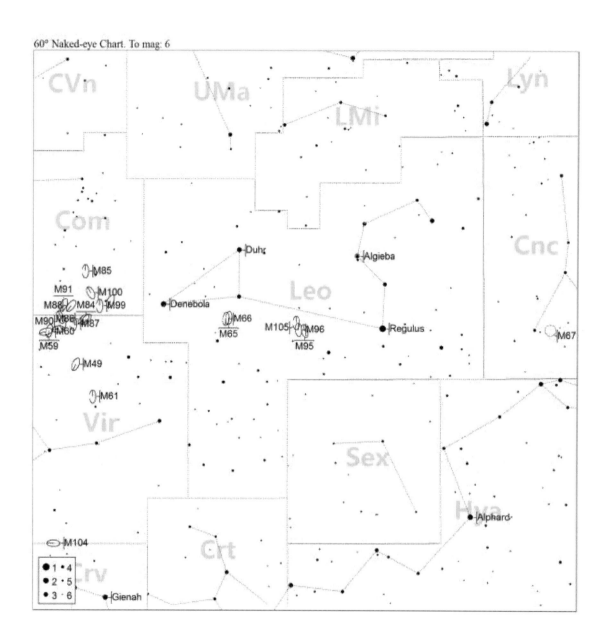

Messier 96:

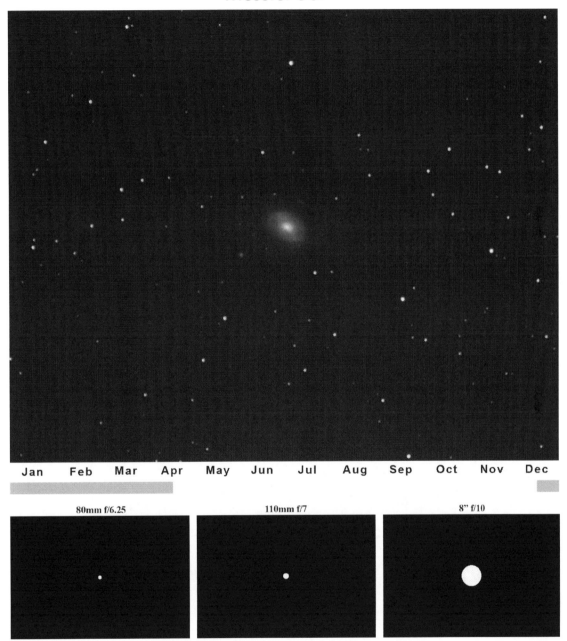

About the target: This intermediate spiral galaxy in the constellation Leo was discovered by Pierre Méchain on the 20th of March, 1781 and subsequentially cataloged by Messier four days later. Sitting some 31 million light years away and spanning some 66,000 light years across, this galaxy is inclined roughly 53 degrees, making it an interesting target if you have the focal length for it. This galaxy has also seen a supernova recently, SN 1998bu, on the 9th of May, 1998.

Where it is:
Right ascension: 10h 46m 45.7s
Declination: +11° 49′ 12″

Imaging the target: This is a very interesting spiral galaxy, at least in the Hubble images. Most amateur astronomer images are a little less so. Still, if you have some time in late winter or early spring, you might want to give this target 30 or more shots at 300-480 seconds at ISO 800 with equipment similar to mine and see what you can get. The image I have included is pretty massively cropped so expect it to be pretty small. My image is also massively under exposed at only 10 300 second images so you should be able to improve on it substantially.

60° Naked-eye Chart. To mag: 6

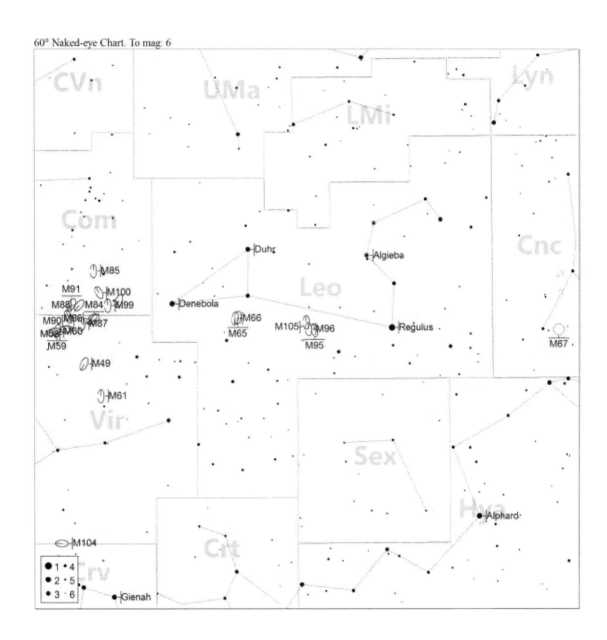

Messier 97: The Owl Nebula

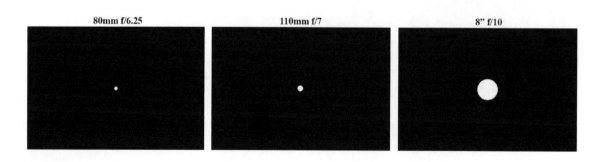

About the target: The Owl nebula is a planetary nebula in the constellation Ursa Major which is an emission nebula ejected by a late term star probably some 6,000 years ago. Discovered on the 16th of February, 1781, by French astronomer Pierre Méchain and then cataloged by Messier on the 24th of March, 1781, it has a diameter of some two light years and is approximately 2,000 light years away. The name came from a drawing done in 1848 by Irish astronomer William Parsons, the 3rd Earl of Rosse where the nebula looked like the head of an owl.

Where it is:
Right ascension: 11h 14m 47.734s
Declination: +55° 01' 08.50"

Imaging the target: Here is another two for one deal with M97 and M108 in the same frame. Since both are very small targets it makes sense for me to include both in one image. With my field of view both targets are too small to get a lot of detail out of so 12 images at 300 seconds at ISO 800 is pretty good, especially since this is an end of winter target so thermal noise is minimal. If you have a longer focal length scope, you may want to include quite a bit more images so you can really get some interesting detail. The background field is pretty sparse so isolating one object from the other for separate stretching is pretty simple. I would suggest separating both objects from the background and processing all three individually so that you keep the colors in the stars and get maximum detail out of all three. That really isn't necessary for monochrome on these targets.

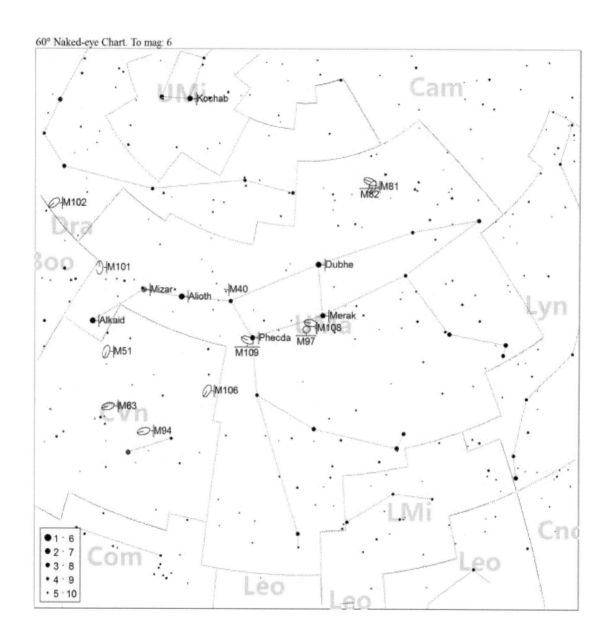

Messier 98:

About the target: This intermediate spiral galaxy was discovered by French astronomer Pierre Méchain on March 15th, 1781 with M99 and M100. It is roughly 60 million light years away and is moving towards us at 140km/sec which makes it one of the few objects moving towards us in our expanding universe.

Where it is:
Right ascension: 12h 13m 48.292s
Declination: +14° 54' 01.69"

Imaging the target: M98 is a spiral galaxy at a pretty severe angle to us and as such, will be very difficult to pull much detail out of. If you pour a lot of time and effort into it, and have a long focal length scope, you can get some nice colors but still, not too much detail. A good place to start with this target will be 300-480 seconds at ISO 800 with equipment similar to mine, and at least 30 or more images to stack. If you really want maximum color details you will need probably 8-10 hours of time with a DSLR or midrange CCD, or maybe 4-5 hours of time on a higher CCD. For me, this target was never really worth it but I have seen some interesting shots by others.

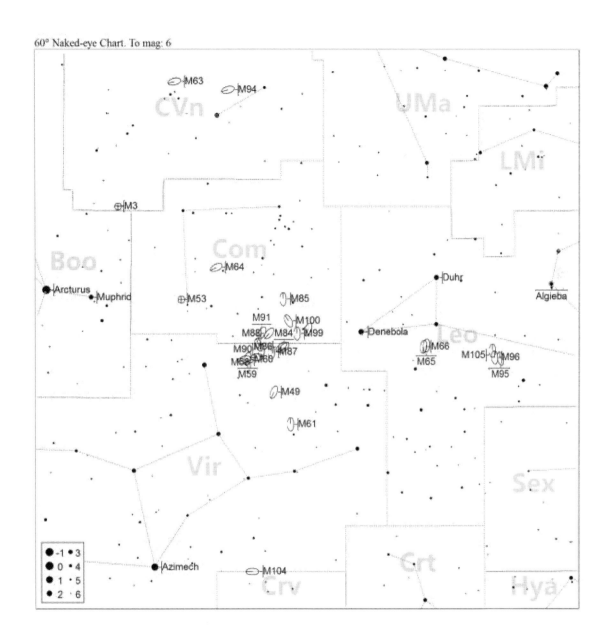

Messier 99: St. Katherine's Wheel

About the target: This unbarred spiral galaxy was discovered by French astronomer Pierre Méchain on the 17th of March, 1781 with M98 and M100. It is roughly 50 million light years away in the constellation Coma Berenices. Messier cataloged the galaxy on the 13th of April, 1781. One of this galaxy's spiral arms is drawn out more than the rest and this could be due to an encounter with another galaxy such as NGC4262 a few hundred million years ago. M99 has been home to three recorded supernovae, the last of which was SN 1986I in May of 1986.

Where it is:
Right ascension: 12h 18m 49.6s
Declination: +14° 24' 59"

Imaging the target: St. Katherine's Wheel is another very small spiral galaxy that can show a little detail with medium focal length scopes but to be really interesting it needs a longer focal length such as with an 8" SCT. This is of course a very deep object where we should start out with 300-480 second exposures at ISO 800 and shoot lots of them, 30 or more to start. Stretching will be a bit of a challenge as well, as this target has several different brightness levels, only two of which are shown in the image I included. With a long focal length scope and a lot of time on the target, this can be a pretty rewarding galaxy.

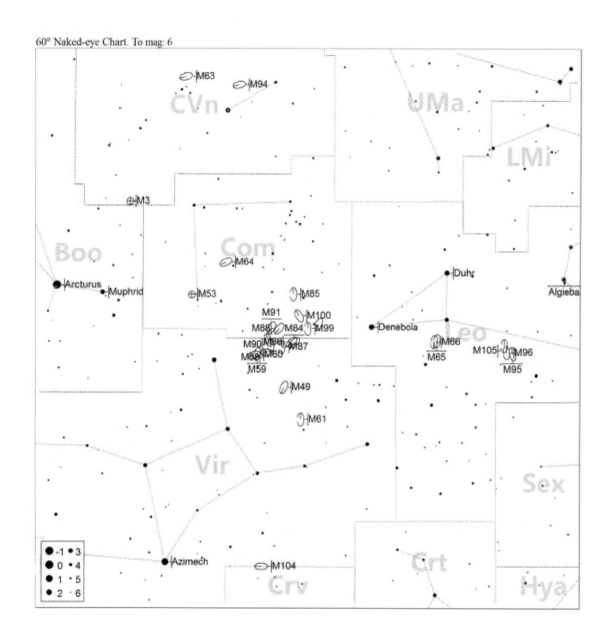

Messier 100:

About the target: This is a prime example of a grand design spiral galaxy sitting roughly 55 million light years away in the constellation Coma Berenices. Over 150,000 light years across it is one of the most visible galaxies in what is called the Virgo SuperCluster, a massive collection of galaxies near the constellation Virgo.

Where it is:
Right ascension: 12h 22m 54.9s
Declination: +15° 49' 21"

Imaging the target: Want another target with free bonuses? Here you go! This spiral galaxy can provide a moderate amount of detail in a medium focal length scope and it can do it while bringing several more galaxies along for the ride. There are approximately 25 other galaxies in the uncropped shot, probably 20 in the shot as is, several of which are pretty obvious. Considering I only shot 10 frames at 300 seconds at ISO 800, that isn't too bad. Increasing to 480 second exposures and shooting at least 30 would have really made not only M100 pop a lot better but would have also showed off the other galaxies as well. Thermal noise is fairly low here as this target is up near the end of winter. I have to make sure this one is on my reshoot list!

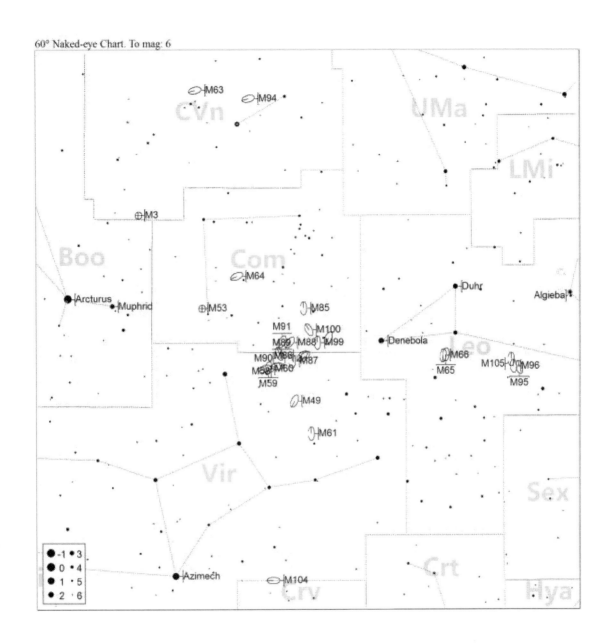

Messier 101: The Pinwheel Galaxy

About the target: Some 21 million light years away in the constellation Ursa Major is the Pinwheel Galaxy, a face on spiral galaxy. This galaxy may be far away but we can see it so well because it is over 165,000 light years across and contains a mass roughly 100 billion times the size of our sun. Note that the top and bottom of the galaxy are not the same, we think this is because in the not too distant past it came very close to another galaxy and the gravitational forces between the two caused the distortions.

Where it is:
Right ascension: 14h 03m 12.6s
Declination: +54° 20' 57"

Imaging the target: The pinwheel galaxy is a nice face on spiral galaxy that, compared to many other Messier galaxies, is relatively easy to capture and process. You will need to go deep to get the details and shoot quite a few frames. The image I included here was shot with 30 frames of 480 seconds each, at ISO 800. I had to do a lot of stretching with curves to get it where it is and have no doubt the level of detail could be increased quite a bit with more light frames. Watch out if you want to mask off the galaxy to preform your stretching as one of the spiral arms is not exactly a spiral but instead just out at an angle.

My image is just beginning to show some of the real detail in the arms. Double the number of exposures and add a little sharpening at the end to a masked off galaxy and it could be a really nice image.

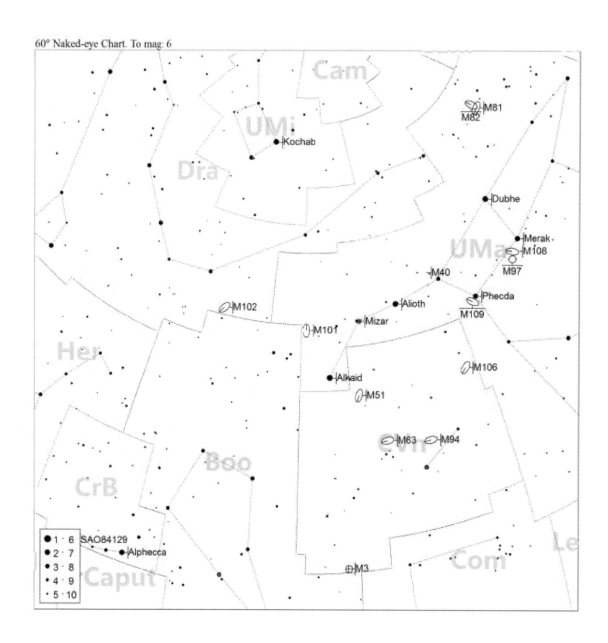

Messier 102:

About the target: This object is the subject of much confusion as it is either a duplication of Meisser 101, or the theory that makes the most sense to me is that it is NGC 5866, a lenticular galaxy in the constellation Draco approximately 50 million light years away. A galaxy matching the description and coordinates (exactly 5 degrees off in RA) of NGC 5866 appears in Charles Messier's notes for March/April 1781.

Where it is:
Right ascension: 15h 06m 29.5s
Declination: +55° 45′ 48″

Imaging the target: This is another of those small, faint galaxies that you see so many of on this list. It also is warm as this one is high in the sky so thermal noise can be an issue. Lots of exposures can really help keep the noise at bay and increase detail although unless you have a longer focal length scope, the detail level will be pretty low. Also try to shoot this one as it is rising early in the morning in late March to get the best temperature; you can start just after midnight around the 15th of the month.

Watch your stretching as well here as the star just up and to the left is pretty bright and can blow out really easily. There are two other pretty bright ones as well to keep an eye on.

60° Naked-eye Chart. To mag: 6

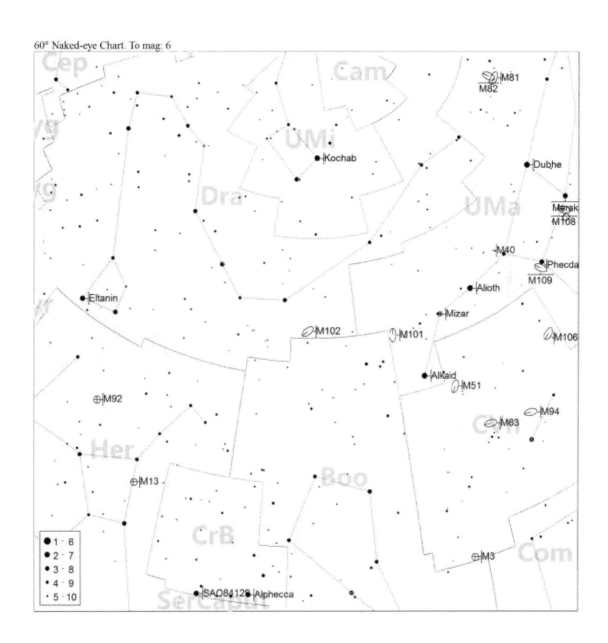

Messier 103:

About the target: About 8,500 light years away in the constellation Cassiopeia lies this open cluster. Approximately forty stars were originally thought to comprise the cluster, however later revisions put the number above 170 spread across 15 light years, and it is thought to be around 25 million years old. This cluster was discovered by French astronomer Pierre Méchain in 1781 and was included in Messier's list even though he did not personally observe it.

Where it is:
Right ascension: 15h 06m 29.5s
Declination: +55° 45′ 48″

Imaging the target: This rather bland open cluster has a fairly compact center and contains some interesting blue and yellow stars which contrast very well. Exposures can be fairly short to medium, 150-240 seconds at ISO 800 for me. You don't need a lot of exposures as thermal noise should be minimal considering the short exposures and that this object is high in the late winter months.

60° Naked-eye Chart. To mag: 6

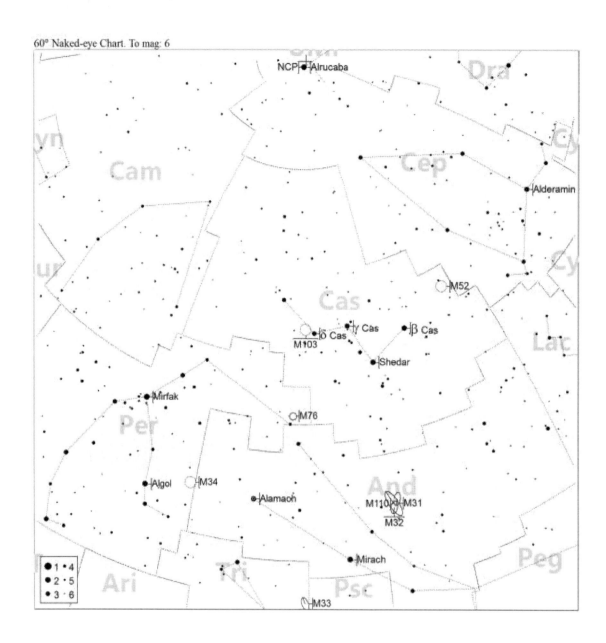

Messier 104: The Sombrero Galaxy

About the target: This unbarred spiral galaxy in the constellation of Virgo is located roughly 28 million light years away and is thought to contain a central supermassive black hole (approximately one billion times the mass of our sun). The dust lane that crosses the central region of the galaxy is what gives it the distinctive look of a sombrero and hence its name.

Where it is:
Right ascension: 12h 39m 59.4s
Declination: −11° 37′ 23″

Imaging the target: The Sombrero Galaxy is a pretty small object, but is extremely well defined. My image had a problem in that the night I was shooting it, the clouds really rolled in so I could only get 10 shots at it. Even so, the galaxy is easily identifiable for exactly what it is. Had I had time I was going for 30-40 shots at 300 seconds at ISO 800 to get the detail level a little higher. When I reshoot I will probably see if I can push it to 480 seconds. Since this is up in the late winter and early spring, thermal noise should really not be too much of a problem even though it will present a small headache given the long exposures and small size of the target.

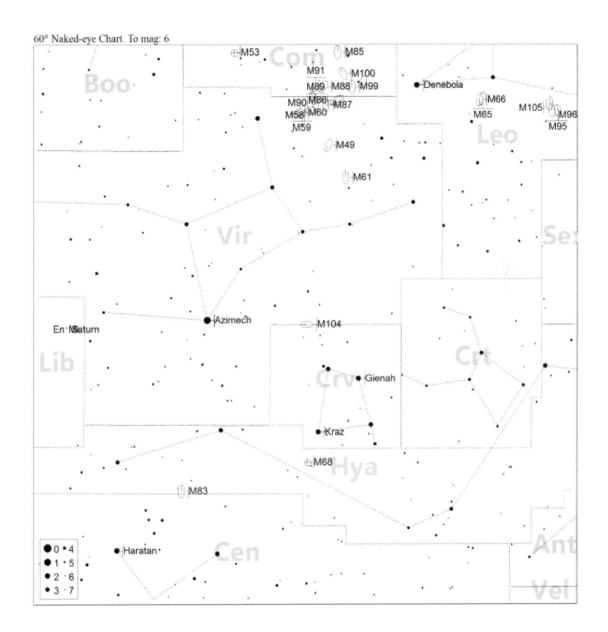

Messier 105:

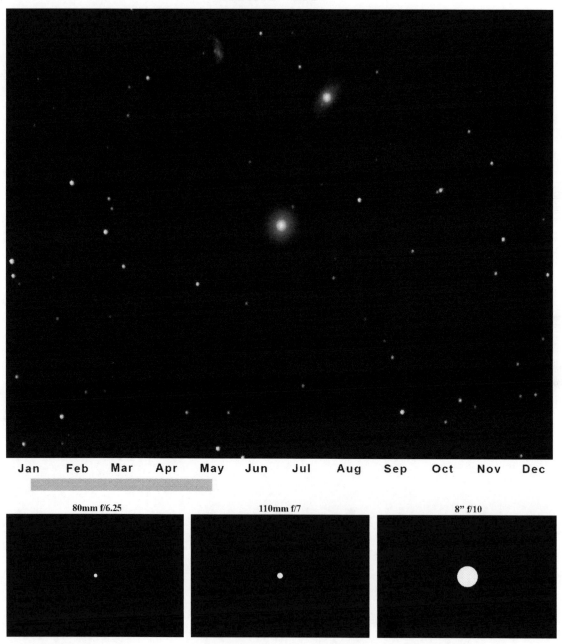

About the target: This simple looking blur is an elliptical galaxy in the constellation Leo and contains a supermassive black hole. Above it and to the right just a little is another elliptical galaxy, NGC3384, to the left of that is yet another galaxy NGC3373, this time a very faint spiral galaxy. These galaxies are part of what is called the Leo SuperCluster. This object as well as M107 and M106 were added to the list by American astronomer Helen Battles Sawyer Hogg in 1947 and may have never been seen by Messier.

Where it is:
Right ascension: 10h 47m 49.6s
Declination: +12° 34' 54"

Imaging the target: Here is yet another fairly boring elliptical galaxy. The saving feature of this image again is that looking closely, you can see several other galaxies as well including NGC3384 and NGC3373. Shooting deep, for me somewhere near 480 seconds at ISO 800 and stretching carefully can make some of these galaxies jump out of this image. In my opinion, the best way to shoot these types of targets is when you can get as many targets in one frame as possible such as the picture I included in the front of this book in the section on the Virgo SuperCluster. There is probably somewhere near 30 galaxies in my field of view of this image but I did not shoot deep enough to really bring them out. There are however three other galaxies in the image clearly visible.

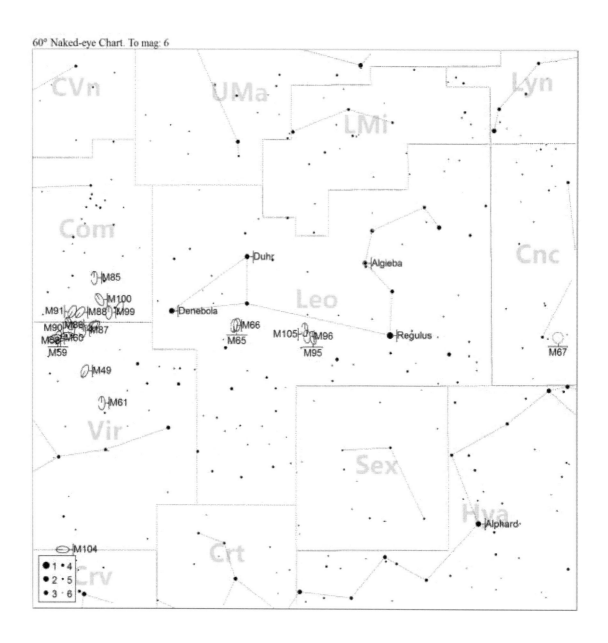

Messier 106:

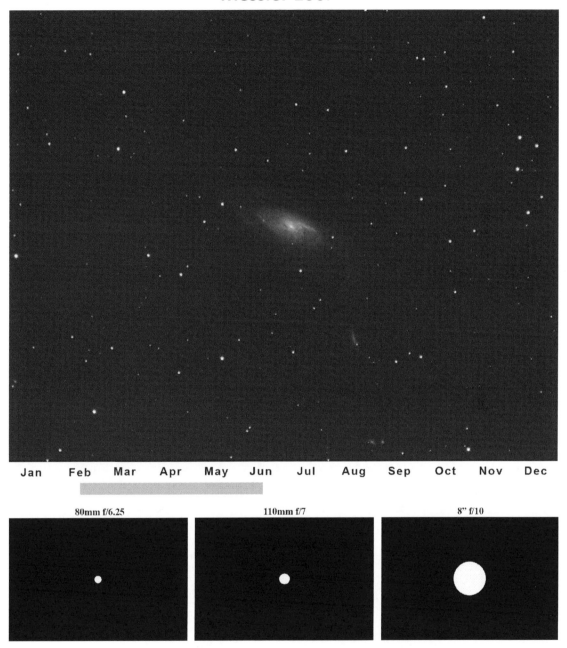

About the target: About 23 million light years away in the constellation Canes Venatici is this spiral galaxy. Down and to the right just a little is NGC4248, another spiral galaxy. If you look really hard you might see a slight fuzzy patch just to the lower left of the main galaxy, it is very faint, but if you do see it (and it is there!) that is PGC39615, another galaxy! This object as well as M105 and M107 were added to the list by American astronomer Helen Battles Sawyer Hogg in 1947 and may have never been seen by Messier.

Where it is:
Right ascension: 12h 18m 57.5s
Declination: +47° 18′ 14″

Imaging the target: Much like M88 and M90, this is an angled spiral galaxy which presents a good level of detail. This one however has the advantage of being up at the end of winter/start of spring so thermal noise will be less of an issue. The image you see is 8 frames of 300 seconds at ISO 1600 and could really use three to four times as much time to really bring out the details. If you look close you can see the grainy texture of the outer arms that resulted from stretching it way beyond what I should have for the amount of time I had on this target. Next time I will most likely shoot 480 second images, 30 or so of them, at ISO 800 to try to keep the noise lower and dynamic range higher for easier stretching.

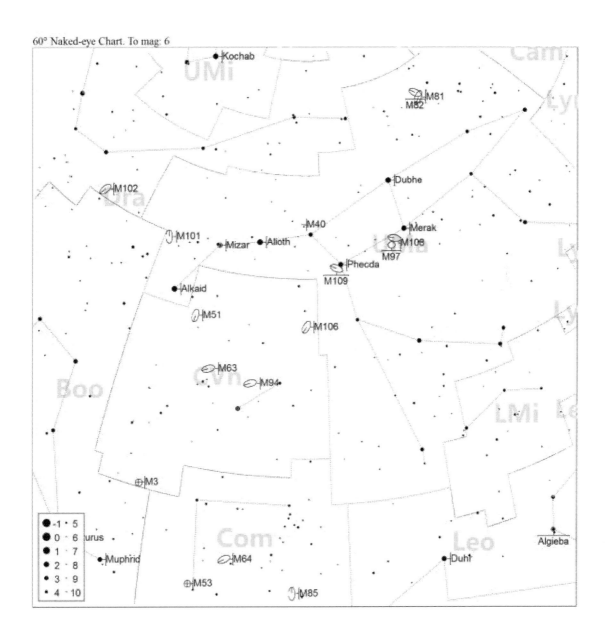

Messier 107:

Jan   Feb   Mar   Apr   May   Jun   Jul   Aug   Sep   Oct   Nov   Dec

80mm f/6.25   110mm f/7   8" f/10

About the target: Discovered in 1786 by French astronomer Pierre Méchain, this very loose globular cluster is some 21,000 light years away and spans almost 80 light years in diameter. This object as well as M105 and M106 were added to the list by American astronomer Helen Battles Sawyer Hogg in 1947 and may have never been seen by Messier.

Where it is:
Right ascension: 16h 32m 31.86s
Declination: −13° 03′ 13.6″

Imaging the target: Another uninteresting globular cluster, but I guess they can't all be M13 now can they. A single set of exposures can capture most of this target but like many globular clusters it could really benefit from having two sets. The first (or only if you just want to shoot one set) would be something like 20 or more medium length exposures around 240 seconds at ISO 800. The second set I would shoot 20 or more frames at 300-480 seconds at ISO 800 to get more of the spread and background field stars. Minimal stretching should provide good star color if you are shooting color. Then I would process these two images in either HDR software or in layers.

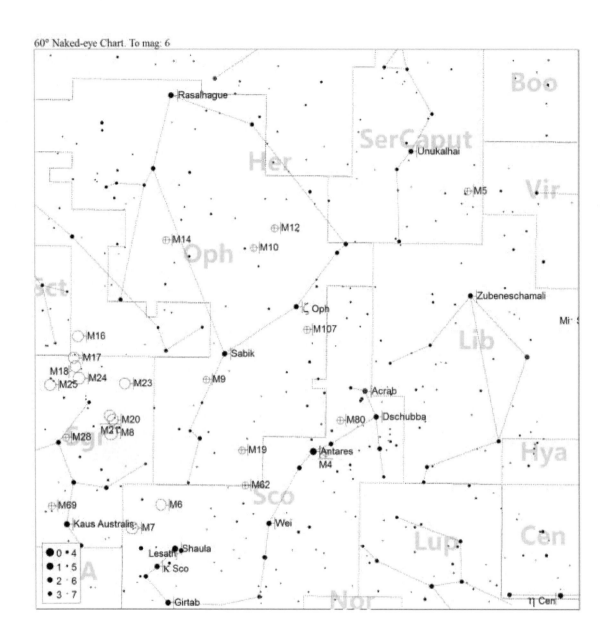

Messier 108:

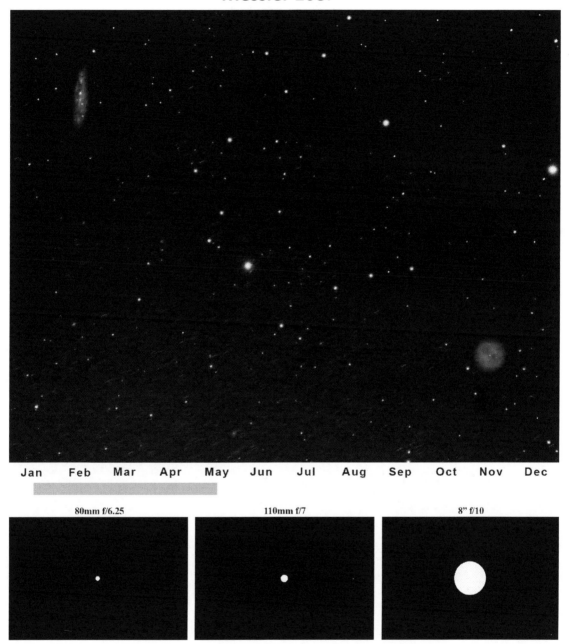

About the target: M108 is in the upper left, M97 the Owl Nebula is in the lower right. The galaxy is a barred spiral galaxy and is in the constellation Ursa Major and contains a supermassive black hole. Discovered by Pierre Méchain around 1781 it was added to Messier's catalog in 1953 by American astronomer and Harvard professor Owen Jay Gingerich. It is thought to be approximately 45,000 light years away and may contain roughly 300 globular clusters.

Where it is:
Right ascension: 11h 11m 31.0s
Declination: +55° 40′ 27″

Imaging the target: Here is another two-for-one deal with M97 and M108 in the same frame. Since both are very small targets, it makes sense for me to include both in one image. With my field of view both targets are too small to get a lot of detail out of so 12 images at 300 seconds at ISO 800 is pretty good, especially since this is an end of winter target, so thermal noise is minimal. If you have a longer focal length scope you may want to include quite a bit more images so you can really get some interesting detail. The background field is pretty sparse so isolating one object from the other for separate stretching is pretty simple. I would suggest separating both objects from the background and processing all three individually so that you keep the colors in the stars and get maximum detail out of all three. That really isn't necessary for monochrome on these targets.

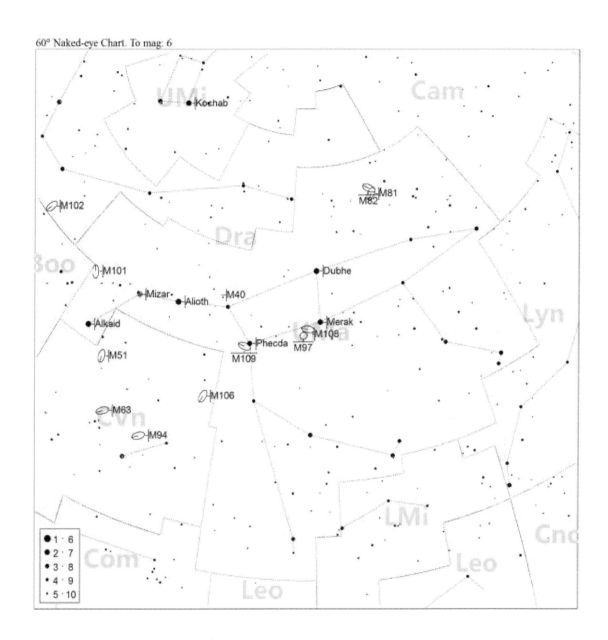

Messier 109:

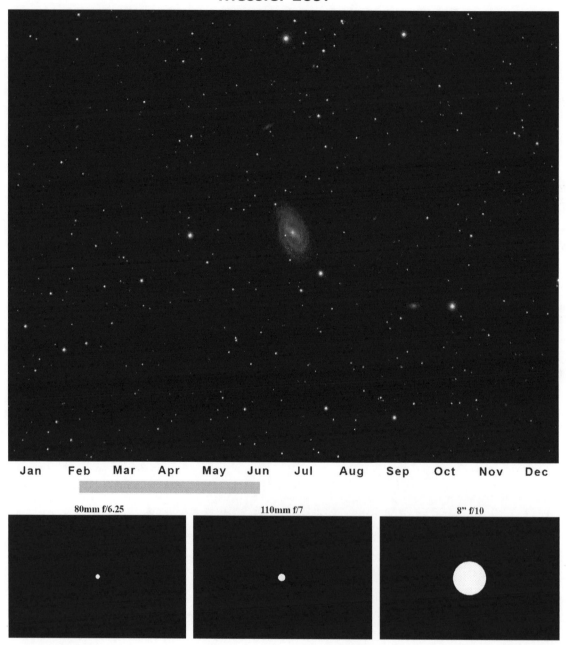

About the target: This barred spiral galaxy in Ursa Major sits some 80 million light years away and was possibly discovered by either French astronomer Pierre Méchain, or Messier in 1781. This galaxy has three known satellite galaxies and has been home to at least one supernova, SN 1956A.

Where it is:
Right ascension: 11h 57m 36.0s
Declination: +53° 22' 28"

Imaging the target: This is a nice example of a barred spiral galaxy almost face on. The slight tilt gives it an interesting appearance. Even with medium focal length equipment you can get some pretty good detail out of it by shooting deep, say 30 or more frames of 300-480 seconds at ISO 800. My shot here is only 10 frames at 300 seconds and still shows a pretty good amount of detail. Like most faint galaxies this is a tough one to stretch requiring a lot of small curve stretches so that the core and surrounding background stars are not completely blown out. Things are just starting to warm a little when this target is up so shoot a few more frames than normal to help combat the thermal noise.

60° Naked-eye Chart. To mag: 6

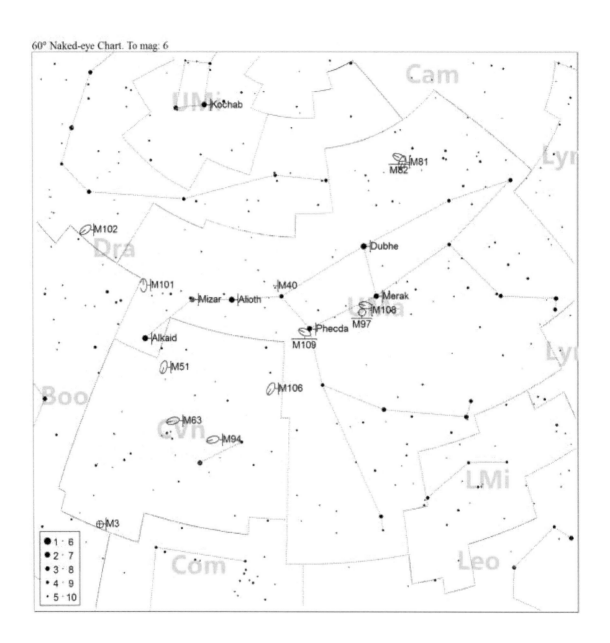

Messier 110:

About the target: This dwarf spheroidal galaxy is a satellite galaxy of the great Andromeda Galaxy (M31). Unfortunately no one really ever pays any attention to this little guy since it is so very close to M31 and in fact, this is simply a crop of M31 to show M110. The only evidence we have that Messier observed this object is that he made a drawing of M31 in 1773 which included this galaxy as well as M32. It was added to Messier's list by British astronomer Kenneth Glyn Jones in 1967.

Where it is:
Right ascension: 00h 40m 22.1s
Declination: +41° 41' 07"

Imaging the target: You have already shot 110, you just didn't know it. There are no "fantastic" images of M110; it is a small satellite galaxy to M31 and you are going to get nothing but a fuzzy blob no matter what you do. Why? Mostly because it is a dwarf elliptical galaxy, which by definition is a small fuzzy blob. The Hubble space telescope took images of M110, care to guess what they looked like? Yep, fuzzy blobs. If this sounds familiar, it should, M31 has two satellite galaxies, both dwarf ellipticals, and both fuzzy blobs.

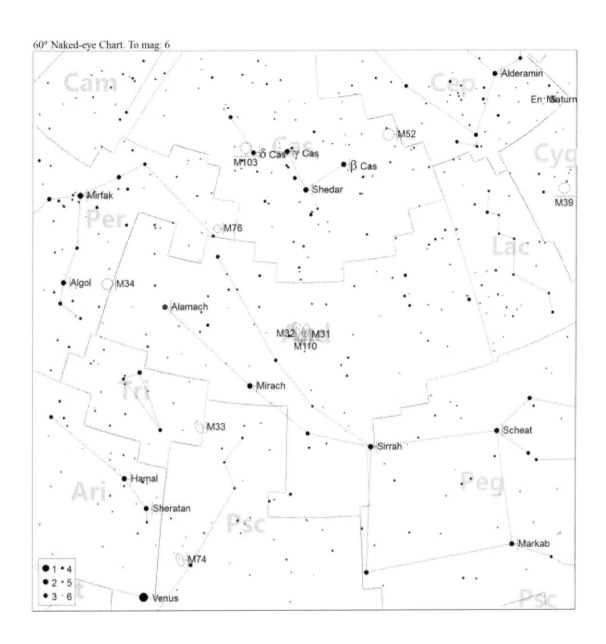

Complete Meisser Shooting Schedule

The following chart shows the best month to image particular Messier objects. The chart was compiled based off target transit times for the 15th day of the month in the central US at as close to midnight as possible.

Jan	Feb	Mar	Apr	May	Jun	Jul	Aug	Sep	Oct	Nov	Dec
44	81	65	40	3	4	6	15	2	31	34	1
46	82	66	49	5	9	7	27	30	32	45	35
47	95	97	53	51	10	8	29	39	33	77	36
48	96	105	58	63	12	11	55		52		37
50		108	59	83	13	14	71		74		38
67			60	101	19	16	72		76		41
93			61	102	62	17	73		103		42
			64		80	18	75		110		43
			68		92	20					78
			84		107	21					79
			85			22					
			86			23					
			87			24					
			88			25					
			89			26					
			90			28					
			91			54					
			94			56					
			98			57					
			99			69					
			100			70					
			104								
			106								
			109								

Closing Notes

I sincerely hope that this book has provided some useful information and made your imaging of the Messier objects a little easier. From here there are many places to go, with my personal favorite being the Caldwell objects.

As always, if you have comments, suggestions, complaints or just want to chat about the book or astrophotography in general check out my website at:

http://www.allans-stuff.com

where you will find more information on astrophotography, my other books, a message board for discussion, videos, and much more.

Good luck and clear skies!

(notes)

(notes)

(notes)

Printed in Great Britain
by Amazon.co.uk, Ltd.,
Marston Gate.